A BROADCAST ENGINEERING TUTORIAL FOR NON-ENGINEERS

Skip Pizzi is Senior Director of New Media Technologies at NAB, where he focuses on new methods for the creation and delivery of broadcast content. He is also Vice-Chair of ATSC *Technology Group 3* (TG3), which is developing standards for the next generation of digital television. Previously he worked in multimedia for 11 years at Microsoft, served as an editor and contributor to several broadcast technology books and journals, and consulted to the professional, educational, and government sectors of the media industry worldwide. He began his career as an engineer, manager, and technical trainer at NPR. He is a recipient of the Audio Engineering Society's Board of Governors Award, and a graduate of Georgetown University, where he studied Electrical Engineering, Fine Arts, and International Economics.

Graham Jones retired in 2010 from NAB, where he was a Senior Director working on advanced television issues. He is still active in ATSC, SCTE, and SMPTE standards committees. Previously he was Engineering Director for the Harris/PBS DTV Express, which introduced DTV to many U.S. broadcasters. He started his career with the BBC in London, and has worked as a consultant to broadcasters in many parts of the world. He holds a degree in Physics and is a chartered electrical engineer, a fellow of SMPTE, and a life member of the SBE and the Royal Television Society. He has been honored with the Bernard J. Lechner Outstanding Contributor Award from the ATSC and received a citation from SMPTE for outstanding service to the society.

A BROADCAST ENGINEERING TUTORIAL FOR NON-ENGINEERS

Fourth Edition

Skip Pizzi

Graham A. Jones

Focal Press
Taylor & Francis Group

NEW YORK AND LONDON

Third edition published 2005 by Focal Press

This edition published 2014
by Focal Press
70 Blanchard Road, Suite 402, Burlington, MA 01803

and by Focal Press
2 Park Square, Milton Park, Abingdon, Oxon OX14 4RN

Focal Press is an imprint of the Taylor & Francis Group, an informa business

Notices
Knowledge and best practice in this field are constantly changing. As new research and experience broaden our understanding, changes in research methods, professional practices, or medical treatment may become necessary.

Practitioners and researchers must always rely on their own experience and knowledge in evaluating and using any information, methods, compounds, or experiments described herein. In using such information or methods they should be mindful of their own safety and the safety of others, including parties for whom they have a professional responsibility.

Product or corporate names may be trademarks or registered trademarks, and are used only for identification and explanation without intent to infringe.

Library of Congress Cataloging in Publication Data
Pizzi, Skip.
A broadcast engineering tutorial for non-engineers / Skip Pizzi, Graham Jones. — 4th edition.
pages cm
Includes index.
1. Radio—Transmitters and transmission. 2. Television—Transmitters and transmission.
3. Radio broadcasting. 4. Television broadcasting. I. Jones, Graham (Electrical engineer) II. Title.
TK6561.P59 2014
621.384—dc23
2013049657

ISBN: 978-0-415-73338-0 (hbk)
ISBN: 978-0-415-73339-7 (pbk)
ISBN: 978-1-315-84842-6 (ebk)

Typeset in Dante MT
By diacriTech, Chennai

Contents

Preface   xi

1 Introduction   1

BROADCASTING BASICS   **3**

2 Types of Broadcasting   5
       Analog Radio   5
       Digital Radio   6
       Satellite Radio   8
       Analog Television   9
       Digital Television   10
       Mobile Digital Television   12
       Cable Television   15
       Satellite Television   16
       Telco Television   18
       IPTV   19
       Internet Radio and Television   19
       Stations, Groups, and Networks   21
3 Sound and Vision   25
       Sound, Audio, and Hearing   25
       Light, Video, and Vision   27
       Baseband   29
4 Analog Color Television   31
       NTSC   31
       PAL and SECAM   39
       HD Analog Video   40
5 Digital Audio and Video   41
       Digital Audio   41

	SD and HD Digital Video	48
	Audio and Video Data Compression	58
6	Information Technology	61
	Binary	61
	Computers	64
	Storage	66
	Computer Networks	68
7	Radio Frequency Waves	75
	Electromagnetic Waves	75
	Frequencies, Bands, and Channels	77
	RF Over Wires and Cables	79
	Modulation	79

STUDIOS, PRODUCTION, AND PLAYOUT FACILITIES   85

8	Radio Studios	87
	Types of Studios	87
	Studio Operations	88
	System Considerations	91
	Audio Mixing Consoles	94
	Microphones	98
	Loudspeakers and Headphones	100
	CD Players	102
	Hard Disk Recorders and Audio Workstations	103
	Radio Program Automation	105
	Digital Record/Playback Devices	106
	Analog Devices	108
	Telephone Hybrids	110
	Remote Sources	111
	Audio Delay Units	111
	Emergency Alert System	112
	Audio Processing Equipment	113
	Signal Distribution	115
	IP-Based Studio Infrastructure ("Audio Over IP")	118
	Ancillary Systems	118
	Radio Master Control	119
	Facilities for IBOC Operations	119
	Radio Data Services	120
	Internet Radio Operations	121
	Other Considerations	122
9	Television Studios and Playout Facilities	123
	Station and Network Operations	123
	Types of Studios	125

Studio Characteristics   125
System Considerations   129
Studio System   130
Post-Production Edit Suites   131
Picture and Waveform Monitoring   134
Television Cameras   136
Film in Television   140
Video Recording   142
Video Editing   152
SMPTE Timecode   152
Video Servers   153
Nonlinear Editing   155
Character Generators and Computer Graphics   156
Electronic Newsroom   157
Signal Distribution   157
Video Timing   159
File-Based Workflows   160
Audio for Television   160
Ancillary Systems   164
Ingest and Conversion   165
IP-Based Studio Infrastructure   166
Television Master Control   167
Television Automation   170
ATSC Encoding and Multiplexing   172
Multicasting Operations   172
Closed Captioning   172
Video Description   173
Alternate Language Audio   173
PSIP Generator   174
Data Broadcasting Equipment   174
Advanced Programming Services   174
Bitstream Distribution and Splicing   176
Signal Delivery to MVPD Headends   177
Internet TV Services   177
10 Remote Broadcasting   179
Radio News Gathering   179
Radio Remote Production   181
Television News Gathering   182
Television Remote Production   184
11 Links   187
Link Architectures   187
Contribution Links for Radio   190
Contribution Links for Television   193

Network Distribution Links for Radio and Television   196
Studio-Transmitter Links for Radio and Television   197

TRANSMISSION STANDARDS AND SYSTEMS   201

12 Analog Radio   203
AM Transmission   203
Emissions Masks   205
FM Transmission   206
Stereo Coding   207
Subcarriers   211

13 IBOC Digital Radio   213
Phased IBOC Introduction   213
Carriers and Channels for IBOC   214
Modulation and Forward Error Correction   214
Audio Data Compression   215
AM IBOC   215
FM IBOC   217
Program and Service Data (PSD)   218
Digital Radio Data Broadcasting   219
Advanced Services   219
HD Radio Standardization   219

14 Alternate Radio Delivery Systems   221
Internet Radio Streaming   221
Audio Podcasting   224
Mobile Radio "APPS"   224
Converged Receivers: "Connected Cars" and Radios in Smartphones   224
Hybrid Radio   226
Audio-Only Service Via DTV   227

15 NTSC Analog Television   229
Carriers and Channels for Analog TV   229
Video Signal   230
Audio Signal   231
Vertical Blanking Interval (VBI) Ancillary Information   232
Closed Captioning and Content Advisory Ratings   232
Analog TV Receiver   233

16 ATSC Digital Television   235
ATSC and the FCC   235
The U.S. Digital TV Transition   236
DTV System   238
Carriers and Channels for DTV   239
8-VSB Modulation   240
ATSC Compressed Bitstream   242

ATSC Video Formats 243
Aspect Ratio Management 244
MPEG-2 Compression 246
Advanced Video Codecs 252
Compression Artifacts 253
AC-3 Audio 253
Advanced Audio Codecs 257
Multiplexing 258
Multicasting 258
Closed Captions 260
Program and System Information Protocol (PSIP) 260
Data Broadcasting and Interactive Television 262
Content Protection (Conditional Access) 265
Advanced ATSC Services 265

17 Alternate Television Delivery Systems 267
Internet Television Streaming and Downloading 267
Connected Television 271
Video Podcasting 272
Mobile Television "APPS" 272
Second Screen and Social TV 272
Hybrid TV 274

18 Next-Generation Broadcast Television Systems 275
Proposed Differences from Current Systems 275
System Proposals around the World 278
New Directions in Audience Measurement 280

19 Transmitter Site Facilities 281
Incoming Feeds 282
Processing Equipment 283
Exciters 284
Power Amplifiers 286
Transmission Lines and Other Equipment 289
AM Antenna Systems 290
FM and TV Antennas 293
Towers 296
Translators and Repeaters 297
Transmitter Remote Control 297
Backup Systems 299

20 Radio Wave Propagation and Broadcast Regulation 301
AM Propagation 301
FM Propagation 304
IBOC Considerations 305
TV VHF and UHF Propagation 306
Spectrum Allocation 307

FCC Rules 308
Spectrum Auctions 312
21 Conclusion 315
Further Information 316

Index 317

Preface

There are many people without engineering backgrounds who need to have a general understanding of broadcast engineering principles. Reaching a large range of devices from TV screens and clock radios to smartphones and digital dashboards, today's broadcasting brings together a wide range of professionals, both technical and non-technical, all working within its vast and omnipresent ecosystem. They may be broadcast managers, program producers, or others who deal with broadcast clients. It is important that they all share some level of knowledge about the workings of broadcast technology.

This tutorial is therefore intended to help such non-engineers seeking to learn about the technology of radio and television. It should also be useful for broadcast engineers in training, or those in technical occupations (such as IT) who find themselves involved with broadcast operations, or who simply want an overview of areas outside their primary expertise. This book explains the jargon of broadcasting and describes the underlying principles, standards, and equipment for broadcast facilities in terms a layperson can understand.

The fourth edition has been completely revised and updated to reflect the increasing use of digital and networking techniques in all aspects of television and radio broadcasting. New chapters have been added to provide an overview of new-media applications by broadcasters and emerging standards in the broadcast industry. The focus is on over-the-air broadcasting from U.S. radio and television stations, but other methods of program delivery to the home are also covered, along with some of the different standards and technologies used in other countries.

Although later chapters build on information in earlier sections, this book can also be consulted for discrete information about a particular topic, and is copiously cross-referenced. However it is used, the overall goal of this book is to help readers further their understanding of the broadcast industry, and thus enhance their ability to perform the broadcast-related functions of their jobs.

Skip Pizzi
NAB Technology Department

xi

CHAPTER 1

Introduction

Broadcasting is a communications service that possesses two fundamental and unique attributes: (1) Broadcasting is a *point-to-multipoint* service, meaning that a broadcast service originates from a single transmitter but is receivable by an unlimited number of receivers within the coverage zone of that transmitter. (Compare this to a *point-to-point* communications system, such as telephony, in which one device typically connects only to a single device at any given time.) (2) Broadcasting is a *unidirectional* service, meaning that it carries content only in a one-way path—from the broadcast station's transmitter to the listener's or viewer's receiver—with no provision for carrying signals back the other way. (Compare this to a *bidirectional* communications system, such as telephony, in which each user's device serves as both a transmitter and a receiver, and content can flow between users in both directions simultaneously.) These two characteristics have defined broadcasting since its origin and continue to do so today. Thus, all the systems described in this book will possess these two essential qualities.

Meanwhile, the *business* of Broadcasting has evolved to include its own two primary functions: (1) The *generation* of audio or audiovisual media content, and (2) the *delivery* of such content to audiences. All broadcasting facilities are organized around one or both of these two processes. Radio and television stations typically include one of each type of facilities. Often these are placed at two different physical locations, but occasionally they are collocated.

Therefore, in its simplest form, a radio or television broadcast station consists of two basic facilities: the studio site and the transmitter site. The studio is where the content is generated. The transmitter site is where the content is sent out over the air, in a point-to-multipoint, unidirectional fashion. If those two facilities are not physically in the same place, between them is a connection called the studio-transmitter link (or "studio-to-transmitter link"), often abbreviated as "STL." But there are many individual components in each of these facilities that make up the chain from content generation to the reception of broadcast services by the viewer or listener. This tutorial provides an introduction to the technologies and equipment that make up these chains, and thereby constitute modern broadcasting systems.

Traditionally, broadcasting was based on *analog* techniques, but for the past quarter-century or so, there has been a steady migration to *digital* systems, which provide many technical and operational benefits for broadcasting processes. The increasing use of computer-based systems has revolutionized both radio and television studios, increasing the quality and efficiency of audio and visual media content creation. More recently, new standards have evolved that also allow application of digital techniques to the transmission of content to end users of both radio and television, improving the efficiency and quality of media content delivery, as well.

All types of broadcast stations used for domestic broadcasting (AM and FM radio, and television) are covered in this tutorial, with descriptions of both analog and digital studio and transmission systems. For completeness, satellite, cable, and Internet delivery are also briefly described, although this book does not cover them in detail.

Chapters in the first section of the book, "Broadcasting Basics," discuss the main methods used for radio and television broadcasting and explain some of the basic science and the terms used later in the book. Chapters in the second section, "Studios, Production, and Playout Facilities," describe radio and television studios and remote operations, covering the main items of equipment used and how they work together. Chapters in the third section, "Transmission Standards and Systems," discuss the standards and technologies used for U.S. radio and television transmission and cover transmitter-site facilities and equipment. The penultimate chapter of this section discusses radio wave propagation and regulation of broadcasting by governmental authorities.

In each section or chapter, audio and radio topics are generally treated first, followed by video and television subjects.

Jargon words and phrases are shown in *italics* when these are used for first time in each section. The words and phrases may be defined in each section or covered in detail in other chapters. Some jargon words are unique to broadcasting, but some are regular words used in a special way that will be explained in each case.

BROADCASTING BASICS

CHAPTER 2

Types of Broadcasting

For many years, the term *broadcasting* meant the transmission of audio or video content via radio-frequency (RF) waves, often referred to as *"over-the-air."* More recently, with developments in advanced digital technology, the term applies to many different types of content distribution. Let's start with a summary of the main types of broadcasting in use today in the United States and elsewhere.

Many of the systems mentioned below differ only in the particular method of transmission or distribution used, whereas the studio systems used for generations of radio and television content have fewer variations. Don't worry if you don't fully understand all of the terms used in this chapter. They will be explained later in the appropriate sections.

ANALOG RADIO

Traditional radio broadcasting for local stations in the United States and throughout the world generally falls into two main types: AM and FM—standing for *amplitude modulation* and *frequency modulation*, respectively. These are the particular methods of radio transmission used for many years in broadcasting audio signals to home, car, and portable receivers. In North America, AM is used in the *medium frequency* (MF)—also known as *medium wave* band—whereas FM uses the *very high-frequency* (VHF) band.

In the United States, a given radio station frequently feeds only one transmitter and therefore is referred to as an AM station or an FM station. It is, however, quite possible for a station to feed both AM and FM transmitters in the same area, or to feed more than one transmitter covering different areas, in which case the term AM or FM may refer only to a particular transmitter and not to the station as a whole. The latter arrangement is more frequently encountered outside the United States, but is becoming increasingly common in the United States.

In some countries, AM radio also uses the *long wave band*, with frequencies somewhat lower than the MF band, and having slightly different propagation characteristics, better for broadcasting over wide areas. AM is also used for *shortwave* radio broadcasting—also known as "HF" for the *high-frequency* band that is used. This is used for broadcasting over very long distances (usually internationally).

We cover analog radio in more detail in Chapter 12.

DIGITAL RADIO

There are four standards for over-the-air digital radio systems in the world, all different from each other in several respects. They are commonly referred to by their acronyms: IBOC, DAB, ISDB-TSB, and DRM.

IBOC (In-Band On-Channel)

Digital radio broadcasting for local stations in the United States uses a system called In-Band On-Channel (IBOC, often pronounced "EYE-bock"). The IBOC digital radio system was developed and continues to be managed by a single company, iBiquity Digital Corporation, referring to its implementation of IBOC by the trademarked name of *HD Radio*. For this reason, the two terms are essentially interchangeable in most practical parlance, and so today, the HD Radio label is more commonly applied to the format. (The HD Radio trade name has led many to assume that the "HD" stands for high definition, but in fact, iBiquity Digital specifies that it is not an acronym and simply an identifier.)

The technology is called In-Band On-Channel because it places a radio station's digital signal within the same band (AM or FM) as the station's original analog system, and within the station's existing analog channel in each case. For this reason, IBOC digital radio does not require any additional spectrum, unlike most other digital broadcasting systems. Today's IBOC station therefore transmits two versions of its primary content—one analog and one digital—thereby serving both legacy and new receivers via the same broadcast channel. (The IBOC system also provides the capability of eliminating the analog component and moving to an all-digital channel, but this mode of operation is not currently allowed by FCC rules).

There are two variants of IBOC: one for AM radio services and one for FM. The major advantage for AM radio is a qualitative improvement in received audio and freedom from the ever-growing impact of audible interference that plagues AM reception. The FM IBOC system also provides better audio quality than traditional analog FM service, but with less noticeable effect since analog FM audio quality is already relatively good to begin with. The primary improvement with FM IBOC is *quantitative*, in that it also allows a station to include multiple audio services within the same broadcast channel (called "multicast services," but often referred to as "HD-2" or "HD-3" services, owing to how the channels are identified on most IBOC receivers). FM IBOC also offers data services that stations can package with each audio service, providing text and other information associated with the audio program, such as song titles and performers' names—so-called "Program and Service Data" (PSD). In addition, FM IBOC can carry advanced data services such as "song tagging" (for identifying content in an online store, from which the listener can download to add the song to a personal music collection) and road-traffic information.

Of course, new HD Radio receivers are required to receive all these digital improvements, but such receivers also operate as analog radios (since both the digital and analog services currently occupy the same radio channels), allowing significant flexibility to broadcasters as they add digital services individually over time.

HD Radio was introduced for regular use in the United States in 2003 and at this writing, there are now more than 2,000 U.S. FM and AM stations carrying IBOC digital radio services, with several other countries considering its adoption. Most U.S. HD Radio stations are in the FM band, and the majority of those currently offer one or more multicast services.

Further detail on IBOC digital radio can be found in Chapter 13.

DAB

Outside the United States, another form of digital radio broadcasting called DAB is in use by some countries. DAB stands for Digital Audio Broadcasting, which is also known as Eureka 147, and in the United Kingdom, simply as Digital Radio. DAB has quality advantages similar to FM IBOC but is fundamentally different in its design. Unlike IBOC, DAB cannot share a channel with an analog broadcast, so new, dedicated spectrum is required. Each individual DAB transmission also requires much more spectrum because it contains multiple program services (typically 6 to 10, depending on quality and the amount of data carried). This makes it impractical for use by the typical local radio station, so it is generally implemented with the cooperation of several broadcasters, or by a third-party aggregator that acts as a transmission service operator for broadcasters.

DAB is most often transmitted using spectrum in the VHF band, which in some countries is becoming available as analog television is phased out. It can also be transmitted in UHF spectrum, or in the *L-Band* (see explanation of Frequencies, Bands, and Channels in Chapter 7).

In recent years, enhanced versions of DAB known as DAB+ and DAB-IP have been developed that increase the capacity of a DAB signal. At this writing, approximately 40 countries around the world have DAB services on air (most are in Europe), and others are considering adoption of DAB or one of its variants. A few countries where DAB is implemented have begun plans to shut down analog radio broadcasting and replace it with DAB service, but the format's success to date has varied widely in different regions. For example, in the United Kingdom, DAB receiver sales now outpace the sale of conventional radios, whereas in Canada, DAB service that the country initiated in 1999 was shut down in 2010 due to lack of consumer adoption.

ISDB-TSB

ISDB-TSB stands for Integrated Services Digital Broadcasting–Terrestrial Sound Broadcasting and is the digital radio system developed for Japan, where the first services started in 2003. Like DAB, ISDB-TSB is intended for multi-program services and is currently using transmission frequencies in the VHF band. One unique feature of this system is that the digital radio channels are intermingled with ISDB digital television channels in the same band.

DRM

DRM stands for Digital Radio Mondiale, a system developed primarily as a direct replacement for AM international broadcasting in the shortwave band, although DRM can also be used in the medium wave and long wave bands. DRM uses the same channel plan as the analog services, and, with some restrictions and changes to the analog service, a DRM broadcast can share the same channel with an analog station. DRM is a monaural (single audio channel) system when used with existing channel allocations, but stereo (two-channel) audio may be possible in the future, if wider broadcast channels are available. DRM started trial implementations in several countries in 2003 and is now in regular use by a number of major broadcasters around the world, primarily in Europe and India. The DRM receiver marketplace remains limited at this writing, however, and consumer uptake has not been particularly robust to date.

DRM+, an enhanced version of DRM, was introduced in 2007. It is designed for use in the VHF band and offers stereo and surround-sound capability. At this writing, DRM+ is undergoing tests in a variety of countries including Germany and Brazil, but it has not been widely implemented anywhere in the world to date.

SATELLITE RADIO

Sirius XM

Sirius XM is the only company that provides a satellite-based digital radio service in the United States. It resulted in 2008 from the combination of two similar but competing satellite radio services: *XM Satellite Radio* and *Sirius Satellite Radio*, both licensed by the FCC as Satellite Digital Audio Radio Services (SDARS). XM and Sirius, which still operate separately at the retail level, are subscription services that broadcast more than 150 digital audio channels intended for reception by car, portable, and fixed receivers. Currently, each service requires two different receiver models and satellite delivery systems, but programming is largely duplicated. The company also has an affiliated operator providing a separate but similar service in Canada known as *Sirius XM Canada*.

XM has four (of which two are spare) high-power *geostationary* satellites (their location in the sky does not change relative to the earth's surface) that transmit frequencies in the *S-Band* (see explanation of Frequencies, Bands, and Channels in Chapter 7). These provide coverage of the complete continental United States, much of Canada, and parts of Mexico. Sirius is similar, except that it uses three *highly elliptical-orbit* (HEO) satellites, with more coverage of Canada, Mexico, and the Caribbean than XM. In 2009, Sirius added a geostationary satellite to enhance coverage further. Both systems use ground-based *repeaters* to fill in many gaps where the satellite signals may be blocked, for example, by large buildings or by tunnels.

1worldspace

1worldspace Satellite Radio (previously known as WorldSpace) is an international satellite radio service that broadcasts about 70 digital audio channels, most by subscription and some free of charge, to many countries around the world. 1worldspace utilizes two geostationary satellites covering Africa, the Middle East, most of Asia, and much of Western Europe. Transmissions are intended for reception by portable and fixed receivers using frequencies in the L-Band. At this writing, the service is reportedly near-defunct.

ANALOG TELEVISION

NTSC

In North America, Japan, and various countries, television was broadcast for many years using the NTSC system. NTSC stands for National Television System Committee, which developed the original standard. The standard defines the format of the *video* that carries the picture information with 525 lines and 30 frames per second, and how the video and audio signals are transmitted. NTSC is broadcast over the air on channels in the VHF and UHF bands. NTSC television can also be carried on analog cable and satellite delivery systems.

In the United States, NTSC is being phased out and has largely been replaced by ATSC digital television. All full-power NTSC transmissions ceased in 2009 in the United States, and NTSC transmissions from U.S. low-power stations and *translators* (a form of repeater station) are scheduled at this writing to be shut down in 2015. NTSC services are also expected to cease in Canada (replaced with ATSC DTV), although the original planned date in 2011 passed without the shutdown taking place, and at this writing, no new date has been set. Mexico is scheduled to end NTSC broadcasting in 2021 (also replaced with ATSC). Japan, Taiwan, and South Korea shut down their NTSC broadcasting services in 2012.

We cover NTSC in more detail in Chapters 4 and 15.

PAL and SECAM

Many countries in Europe, Australia, and other parts of the world use an analog color television system called PAL. The underlying technologies used are the same as NTSC, but the color coding system and picture structure are slightly different, with 625 lines and 25 frames per second. PAL stands for Phase Alternating Line, which refers to the way the color information is carried on alternating lines. SECAM is another color television system used for transmission in France, Russia, and a few other countries. SECAM stands for the French words Sequential Couleur avec Mémoire, which refer to the way the color information is sent sequentially and stored from one line to the next. PAL television signals are transmitted in a similar way to NTSC, but the bandwidth of the RF channel is different; the SECAM transmission system has several differences from both NTSC and PAL.

PAL and SECAM television is being gradually phased out in favor of digital television. Most European countries have already ceased PAL transmissions, and many other PAL countries around the world are planning to complete their transition to digital between 2015 and 2020.

DIGITAL TELEVISION

Over-the-air digital television, DTV, is also referred to as Digital Terrestrial Television Broadcasting, or DTTB. At this writing, there are four main DTV systems in the world, with significant technical differences: ATSC, DVB-T, ISDB-T, and DTMB.

ATSC

ATSC stands for Advanced Television Systems Committee and is the DTV standard for the United States, where DTV broadcasting started in 1996. ATSC has also been adopted by Canada, Mexico, Korea, Honduras, and El Salvador, and is being considered by some other countries. The ATSC system allows transmission of both *standard definition* (SD) and *high-definition* (HD or HDTV) program services using special coders/decoders (referred to as *codecs*) to fit the video and audio signals into the available *bandwidth*. Capabilities include widescreen 16:9 *aspect ratio* pictures, *surround sound* audio, *electronic program guide, multicasting,* and *datacasting.* ATSC DTV is transmitted over the air in the same VHF and UHF bands as NTSC television but with a modulation system called 8-VSB, using vacant channels in the NTSC channel allocation plan for the country. Cable television systems also carry DTV programs produced for ATSC transmission, but do not actually use the transmission part of the ATSC standard in their delivery services.

We cover ATSC and the digital TV transition in more detail in Chapters 5 and 16.

DVB-T

Terrestrial DTV in Europe, Australia, and many other countries uses the DVB-T standard, which stands for Digital Video Broadcasting–Terrestrial. DVB-T allows transmission of both SD and HD programs, and most of its capabilities are generally similar to ATSC. Initially, most countries using DVB-T, apart from Australia, transmitted only in standard definition, but many countries around the world are now introducing terrestrial HD services.

Like ATSC, DVB-T is transmitted over the air in the VHF and UHF television bands. The main difference from ATSC, apart from the details of the picture formats used, is in the technical method of transmission, in which DVB-T uses *orthogonal frequency-division multiplexing* (OFDM). An enhanced version of the standard, DVB-T2, was introduced in 2008 using more efficient, advanced modulation and coding techniques, providing a significant increase in capacity of the channel.

As with ATSC in the United States, DVB-T (or DVB-T2) services will eventually replace analog television broadcasting in the countries where it is used (and in some cases, it has already done so).

ISDB-T

Japan uses ISDB-T, the Integrated Services Digital Broadcasting–Terrestrial standard, to broadcast both SD and HD programs in the VHF and UHF television bands. Modulation and other transmission arrangements have some similarities to DVB-T, but the system uses *data segments* in the transmitted signal to provide more flexible, multi-program arrangements for different services and reception conditions. ISDB-T will eventually replace analog television broadcasting in the countries where it is used (and in some cases, it has already done so).

A variant of ISDB-T known as SBTVD (Sistema Brasileiro de Televisão Digital) has been adopted by Brazil, Argentina, and Peru and, at this writing, is expected to be adopted by other South American countries also. The main variation is in the use of more advanced high-efficiency video and audio codecs compared to the older ISDB-T standard.

Unlike ATSC and DVB formats, the ISDB standard is not named after the organization that created it, which in ISDB's case is the Association of Radio Industries and Businesses (ARIB), a public service corporation chartered by the Japanese government in 1995.

DTMB

China was the last developed country of the world to introduce DTV, using a locally developed standard called DTMB, which stands for Digital Terrestrial Multimedia Broadcast. This system has similarities to both the ATSC and DVB-T systems (but with various differences), using *quadrature amplitude modulation* (QAM), supporting both single and multiple carrier modes, and with a different version of OFDM from DVB-T. DTMB is capable of working with HDTV receivers in vehicles moving at speeds up to 125 mph and also supports mobile digital TV services (see below).

DTMB has been adopted for use throughout the People's Republic of China. Broadcasts began in selected areas during 2008. At this writing, the system is also currently undergoing tests in a variety of other countries in the Middle East and South America.

MOBILE DIGITAL TELEVISION

Mobile digital television systems are intended for transmission of video, audio, and data to receivers in handheld devices such as cell phones and portable media players, and to mobile receivers in vehicles. The main requirements are reduced power consumption to increase battery life in portable devices and a robust signal for reception in difficult conditions. This industry sector has developed very rapidly, and numerous standards and proprietary systems have been introduced in different parts of the world with varying degrees of success. Continuing developments are almost certain. At this writing, there are several mobile DTV systems in use in different parts of the world, in particular: ATSC Mobile DTV, DVB-H, ISDB 1seg, DMB, and CMMB. These have significant technical differences, though for efficient use of bandwidth, they all support advanced video and audio compression codecs, and they all support some form of encryption or conditional access to prevent unauthorized viewing. Video services also can be carried over mobile phone networks that are not covered in this tutorial (since they are not carried by a broadcast service, but rather over a point-to-point wireless data service).

ATSC Mobile DTV

The ATSC Mobile DTV standard was finalized in 2009 and is rapidly being adopted by many U.S. DTV broadcasters for transmissions to handheld and mobile receivers. Initial program offerings have been free to air. The service is "in-band," meaning broadcasters provide the mobile TV services as part of their terrestrial transmission in the same TV channel they use for regular DTV programming. To make space for the Mobile DTV signal, the station gives up part of the bandwidth it may have previously allocated to multicast SD program channels, or perhaps reduces the bandwidth allocated for an HD channel.

Because it uses the existing DTV infrastructure, ATSC Mobile can be added to a DTV station at comparatively low cost.

Video resolution is 240 lines × 420 pixels per line, 16:9 aspect ratio, and an enhanced mode enables higher quality 480p 16:9 SD video to be carried. Depending on the bandwidth allocated, multiple TV and/or audio-only programs are possible, with support for an *electronic service guide*, and *interactive services*. There is also the possibility for content to be downloaded to the receiving device in *non-real time* (NRT) for storage and playback later, and for content to be protected from free access, thus enabling subscription business models.

We cover ATSC Mobile DTV in more detail in Chapter 16.

MediaFLO

MediaFLO (also known as FLO TV) was developed by Qualcomm to transmit TV, audio, and data primarily to mobile phones. The service was considered by several wireless telecommunications companies in the United States, and initially offered by some, but in 2011, MediaFLO service was terminated, ending further development and deployment of the service. Although it targeted phone-based receivers, the system provided transmission of content from a transmitter to receiver in a one-way mode, and hence the name "FLO"— Forward Link Only. MediaFLO in the United States was transmitted by a network of high-power broadcast transmitters that had previously been operated as broadcast UHF TV channels 55 and 56. Transmissions were encrypted and used OFDM with QAM or *quadrature phase-shift keying* (QPSK) modulation. Video resolution was 240 lines × 320 pixels, which is known as the *quarter video graphics array* standard (QVGA).

Verizon's VCAST-TV service was launched in 2007 as a subscription service using MediaFLO technology. AT&T launched a similar service in 2008. Up to 12 channels of entertainment and information programs were carried on these services, but low consumer uptake was cited for both services' termination, followed by Qualcomm's shuttering of its MediaFLO operations. Prior to the shutdown, trials also had been conducted by Qualcomm in various countries overseas. Following the termination of service, AT&T acquired the spectrum in which MediaFLO had operated from Qualcomm.

DVB-H

DVB-H stands for Digital Video Broadcasting–Handheld and is a variant of the DVB-T standard, intended for battery-powered, handheld devices. Transmissions are in the UHF band using OFDM with QPSK or QAM. Video is QVGA resolution, and the number of program channels depends on the bandwidth allocated. Although, in theory, DVB-H services can be carried by a broadcast station along with DVB-T services in a single DTV channel, this has not turned out to be practical due to different reception considerations, and separate transmission networks for DVB-H services were generally implemented. Both free-to-air and subscription business models could be accommodated.

Although DVB-H services were launched in a number of European, Asian, and other countries, consumer uptake was slow, and receiver availability was limited. As a result, many of these services have shut down. A variant known as DVB-SH (Satellite services to Handhelds) was subsequently introduced, which allows for hybrid systems using both satellite and terrestrial transmissions. At this writing, the system is under trial in a few countries, but no commercial service has been introduced yet.

More recently, a successor format called DVB-NGH (Next-Generation Handheld) has been standardized, intended for terrestrial mobile TV use with several key improvements over DVB-H. Trials are also underway at this writing, with initial consumer products not yet available.

ISDB 1seg

ISDB 1seg stands for Integrated Services Digital Broadcasting—One Segment, and is a variant of the ISDB-T standard. It is widely used in Japan for mobile terrestrial digital video, audio, and data broadcasting services, where it was introduced in 2005. It is also used in Brazil, where it was introduced in 2007 in a few cities. It gets its name from the 1 segment that is used for mobile services out of the 13 segments in the ISDB broadcast channel.

Transmissions are in the UHF band and use QPSK modulation. Video is QVGA. There is a data capacity allowed for an electronic program guide (EPG) and interactive services. Conditional access and copy control are implemented by the use of a *Copy Control Descriptor* within the broadcast. The broadcast content itself is not encrypted, but Copy Control information carried in the signal forces the receiver to encrypt stored recordings and disallows users from making a copy of the recording.

DMB

DMB stands for Digital Multimedia Broadcasting, a technology for sending TV, radio, and data to mobile phones and other devices. It is a television adaptation of the DAB digital radio standard described above. The system is in widespread use in South Korea, where it was introduced in 2005. At this writing, it is also available or undergoing trials in various other countries in Europe and elsewhere. It can operate via satellite (S-DMB) or terrestrial (T-DMB) transmission. Spectrum used is in the VHF or L-Band for terrestrial use, and in the S-Band for satellite, using COFDM with QPSK modulation. Video is QVGA. In South Korea, the T-DMB service is provided free of charge to the viewer, being supported by advertising.

CMMB

CMMB stands for China Mobile Multimedia Broadcasting. It is a mobile DTV and multimedia standard used in China, developed by the Chinese government's State Administration of Radio, Film, and Television (SARFT). It is somewhat similar to DVB-SH in that it includes broadcasts from both satellites and terrestrial repeaters to handheld devices. It provides 25 video and 30 radio channels, plus additional data channels in either 2 MHz or 8 MHz bandwidth channels, within 25 MHz in the S-band (2635–2660 MHz). It provides video resolution of *quarter common intermediate format* (QCIF, 176x144) or QVGA and supports reception at speeds up to 155 mph (250 km/h).

CABLE TELEVISION

Cable television operators are referred to as Multichannel Video Programming Distributors (MVPDs), or as *multiple systems operators* (MSOs), if they own systems in more than one community. They distribute large numbers of television and audio program channels from a *cable headend* to consumers over networks of cables spanning urban and suburban areas. They do not usually cover large rural areas due to the greater distances between homes. Such services carry a subscription fee and always carry program services from all or most of the broadcast stations in the area, as well as numerous other channels. Most cable systems provide *video-on-demand* (VOD) and *pay-per-view* (PPV) services as well as regular continuous programming, and use some types of *encryption* or *conditional access* to prevent piracy and unauthorized viewing of some or all of their channels.

Analog Cable

Traditional analog cable carries television channels at RF on one or two *coaxial cables* connected into the home, using similar bands as over-the-air broadcast television, but with slightly different channels and a wider range of frequencies. In the United States, apart from the channel allocations, the analog cable television signal is basically identical to NTSC broadcast over the air. Many cable systems have added digital service, but most of these have retained some level of analog service for legacy customers and hardware.

Digital Cable

Digital cable services can be carried on the same cable as analog, using different channel allocations for the analog and digital signals. To accommodate the additional bandwidth that is required to carry both digital and analog channels, many cable operators have upgraded parts of their distribution networks to optical fiber, but the connection to the home may still be by coaxial cable. In the United States, digital cable may carry SD and HD DTV programs produced for ATSC transmission using MPEG-2 video compression, but the cable modulation system used is QAM, which is different from the over-the-air 8-VSB standard. To accommodate this, ATSC broadcast channels undergo some form of conversion at cable headends, and this may introduce some degree of quality degradation. In 2009, some MSOs started to introduce services using the MPEG-4 *Advanced Video Coding* (AVC) video codec, for more efficient use of the cable capacity. This process implies that ATSC broadcast channels may also undergo *transcoding*, by which they are compressed by two different codecs, creating the possibility of further quality loss.

Digital cable systems have a built-in *return channel* capability over the cable from the consumer to the cable operator. In 2009, some MSOs started to introduce *interactive services* based on a system known as tru2way™, or a simplified version called EBIF (Enhanced TV Binary Interchange Format).

Digital cable systems are also able to offer broadband Internet connectivity to customers over their networks, usually through a standard known as *DOCSIS* (Data Over Cable Service Interface Specification). Many digital cable operators also offer *Voice over IP* (VoIP) telephony services.

Operators choosing to provide all of these services offer consumers so-called "triple-play" services—broadband Internet, voice telephone, and television channels with associated interactive services—all carried on the same cable. Some operators are also incorporating so-called "Switched Digital Video" (SDV) and IP-based services (see below) to add efficiency to their TV offerings.

Both DVB and ISDB have digital cable variants of their DTV standards. Services in Europe and some other countries use the DVB-C (Digital Video Broadcasting–Cable) standard, and Japan uses ISDB-C (Integrated Services Digital Broadcasting–Cable).

SATELLITE TELEVISION

Medium- and Low-Power Services

Many television services are distributed in the United States and elsewhere using medium- and low-power geostationary satellites in the C and Ku-Bands (explained in Chapter 7) some with analog, but most now with digital transmissions. Although most of these are intended for professional users (for example, distribution to cable television headends or to broadcast stations), some can be received by consumers using large satellite dishes, typically 6 to 10 feet in diameter. These are sometimes known as "big ugly dishes" (BUDs) or TV Receive-only (TVRO) terminals. Some channels are transmitted *in the clear* and are free of charge, whereas others are *encrypted* and require a subscription to allow them to be viewed. Many network feeds are carried, but very few individual broadcast stations have their programs distributed in this way.

Digital Satellite Broadcasting

In the United States, *direct broadcast satellite* (DBS) delivery of digital television service to the home, also known as *direct-to-home* (DTH), has been provided by several operators, although currently there are just two competitors: DirecTV and Dish Network. Like cable operators, these are referred to as MVPDs. Unlike cable, however, DBS service is just as easy to provide in rural areas as it is in urban locations, and therefore a high percentage of DBS users are in rural areas not well served by cable (or even terrestrial) broadcast services.

DBS services in the United States use several high-power geostationary satellites to provide hundreds of subscription channels, both SD and HD, including local broadcast station channels in each market area they cover. (In rural areas outside a designated broadcast market, DBS services carry a selected broadcast station from another market for each network, so that these rural users are still able to receive local broadcast content from elsewhere.) These DBS services transmit in the Ku band, and their higher power satellite transmission allows the use of small satellite receive antennas, typically of less than 1 m (39 in.) diameter. (In the continental United States, these dishes are usually 24 inches or less in diameter.) DBS television services provide VOD and PPV, as well as continuous programming, and content is encrypted to prevent unauthorized viewing.

Dish Network uses the DVB-S (Digital Video Broadcasting–Satellite) standard with QPSK modulation, while DirecTV uses its own, slightly different transmission system, also based on QPSK.

There are numerous DBS service providers in other parts of the world. Key features are the capability to cover very large areas—many countries or even a continent—from one satellite, generally with capacity for large numbers of program channels. Services in Europe and many other countries use the DVB-S standard. There is also an enhanced version of the standard, known as DVB-S2, introduced in 2004, utilizing the AVC advanced codec and using 8-level phase-shift keying (8PSK) modulation and a more efficient error-correcting code to provide improved performance and increased channel capacity.

In 2005 and 2006, the US Dish Network and DirecTV adopted the DVB-S2 standard for new DBS services of HD local programming and for some VOD services.

In Japan, digital DBS services use the ISDB-S (Integrated Services Digital Broadcasting–Satellite) standard, which can use various forms of PSK modulation.

TELCO TELEVISION

Digital Optical Fiber

In 2005, the telephone company ("telco") Verizon introduced its *FiOS* Internet, telephone, and television services in selected U.S. markets, thus becoming another MVPD. At this writing, FiOS is now available to about 10 million homes, with more systems planned, but many parts of the country are not covered. The FiOS TV service is similar to digital cable and provides many SD and HD program choices available by subscription, including local station channels, pay-per view, VOD, and interactive capabilities.

Apart from being operated by a telco, the main difference from digital cable is that the FiOS signal are carried over a *FTTP* (Fiber to the Premises) network (also known as *FTTH*, for Fiber to the Home) all the way from the FiOS headend to the subscriber's premises on optical fiber. This provides much greater bandwidth than available on copper coaxial cables and allows the possibility of more television channels and higher-speed Internet connections for consumers.

AT&T introduced another fiber-based telco TV system called *U-verse* in 2006. At this writing, it is now available to approximately 25 million homes, with more systems planned, but many parts of the country also are not covered. U-Verse has some similarities to FiOS, but there are fundamental differences. In most U-verse systems, the optical fiber is used for FTTN (Fiber to the Node), with a *node* point in each neighborhood. The connection from the node to the home is made over twisted pair copper wiring, which limits the bandwidth available for Internet services, and restricts the number of TV programs that can be received simultaneously for recording or viewing on multiple televisions. FTTP is, however, being installed by AT&T in some areas, which is less restrictive. Another big difference for U-verse TV compared to FiOS and digital cable is that it exclusively uses IPTV technology (see later section) and AVC video coding.

More recently, Google introduced Google TV, which is a FTTP service similar to Verizon's FiOS, but offering even higher connectivity speeds for its Internet connectivity service. At this writing, Google TV is available only in few selected markets around the United States.

Digital Subscriber Line

Telcos can also deliver limited television services over *digital subscriber lines* (DSL), which use copper wire rather than optical fiber for the connection from the telco headend to the customer's premises. The level of service, and whether HD or only SD programming is supported, depends on local circumstances and the type of DSL being used. Such services may use IPTV technology, although this is not always the case.

IPTV

IPTV stands for *Internet protocol television*. It is most often used to describe delivery of television programming by a telephone company (telco) to consumers using IP protocol over a private data network. As noted above, AT&T's U-Verse service is a national offering using IPTV, but other operators (typically local or regional telcos) also offer IPTV service. Like most other digital TV systems, IPTV needs a *set-top box* to convert it into standard video signals that can be fed to a regular television. IPTV can be used to deliver either continuous streams of content or video on demand, and it can also be used for non-real-time download to a storage device. Such services are usually based on subscription.

As explained in Chapter 6, IP is a widely used method for directing *packets* of data between devices connected to a network, which may or may not be connected to the Internet. IPTV is really a technology rather than a distribution medium, and it can be delivered over a telephone DSL, optical fiber, digital cable, or wireless transmission. As mentioned, IPTV services presented by MVPDs are generally *not* delivered over the public Internet, but over private IP paths. In addition to its use by telcos, IPTV technology is increasingly used by satellite and cable operators and even by terrestrial broadcast channels as an alternative method of delivering continuous streams or VOD content.

INTERNET RADIO AND TELEVISION

With the rise of the Internet in the 1990s, and the deployment of higher-speed ("broadband") connectivity in the 2000s, a new distribution medium for radio and television programming developed. IP-based streaming and download technologies make possible the distribution of audio and video over the public Internet. Unlike over-the-air broadcasting, the programming is available only to those with access to the Internet over a broadband connection.

Receiving Internet radio and TV originally required the use of computer equipment and web browser software. But stand-alone Internet radio receivers that provide access to Internet radio stations without a computer or a separate audio system are now available. For video, dedicated Internet TV set-top boxes that provide the interface to the Internet, storage, and ability to download or stream video content for playing on a regular television are also available. Some of these devices are designed for use with particular sources of content, while others work with Internet TV content in general. A new generation of integrated television receivers is also now available with Internet capability and, in some cases, data storage, built-in. Such devices are sometimes referred to as "Connected TVs" or "Smart TVs."

In addition, where enabled by the content owner, file transfer capabilities can provide automatic, "push" download of complete programs or program segments for storing and later playback on a computer or other device, including portable players such as the Apple iPod—giving rise to the new term of *podcasting*.

"Over the Top" (OTT) Service

In contrast to the Telco TV and IPTV services described above, a number of television content providers offer on-demand, IP-based television service over the public Internet. Because these services deliver their content via regular Internet connectivity, they are called *Over the Top* (OTT) services—in other words, they utilize an existing delivery system provided by others, rather than establishing their own delivery paths as broadcast, cable, satellite, and telco TV services do.

Users typically access OTT services via one of the many general Internet TV set-top boxes, which are inexpensive third-party devices, or on Connected TVs and Blu-ray players. Here again, OTT services rely on others to provide necessary hardware. Of course, OTT content can also be directly viewed on Internet-connected computers, tablets, and smartphones, via Internet browsers and applications.

OTT services provide either streaming or downloadable content on demand, and are typically subscription based. In some cases, a pay-per-view fee is assessed instead, or in addition to a baseline subscription fee. Users initiate service by downloading an application from the OTT provider to their viewing devices (or in some cases, the application is natively provided on the device), then browse content offerings and order content for immediate streaming views, or as downloads for later viewing. Often, once content is purchased, it can be viewed multiple times, and/or transported across devices and stored for viewing in multiple environments. In this way, OTT services are directly competitive with VOD service from MVPDs.

Service Implications

One element of content not available from OTT services at this writing is live terrestrial broadcast TV service. This has led to a trend called "cord cutting," whereby a growing number of U.S. consumers have moved to over-the-air reception of local broadcast TV, and use OTT for everything else, thereby avoiding the higher monthly cost of MVPD service. (For many consumers, the cost of broadband Internet service plus an OTT subscription comes to less than their monthly MVPD service fees.)

Meanwhile, large numbers of broadcast networks, stations, and other organizations worldwide are now making radio and TV content available on the Internet, including much of their regular broadcast programming. This may be for on-demand delayed playback—for example, a program is made available on-demand one day after it airs over broadcast—or, in some cases, streamed as a simulcast with the original broadcast. (Much of the content available on OTT services is such episodic TV material.)

Outside the United States, the BBC makes virtually all of its radio and television programming available on the Internet through its "iPlayer" website, both live and on-demand, although, for copyright reasons, most of the television content is restricted to Internet addresses within the United Kingdom. Similarly, most U.S.-based OTT television services are limited to U.S. distribution.

Many U.S. radio stations stream all or most of their content live to the Internet at this writing. The exceptions occur when either a station chooses not to stream their service online for business reasons, or when it airs content from a network or other outside provider for which the station has only limited distribution rights (e.g., permission for on-air broadcast but not on-line streaming). This applies to television stations as well, but because the majority of content aired by a local TV station is acquired from other parties (unlike most radio stations), and most rights agreements for this content apply only to on-air broadcast, U.S. TV stations typically do *not* provide full-time, live streaming of their broadcasts, since most of the content aired cannot be legally streamed online.

Whether and how local terrestrial TV broadcasters provide live streaming services via the Internet is the subject of much ongoing negotiation at this writing, however. Such transmissions would require many decisions and changes in existing agreements. For example, most U.S. TV stations' network affiliations grant them exclusive permission to distribute content over a limited geographic area, which is a natural function of broadcast signal coverage. MVPD coverage agreements essentially duplicate such regional limitations, whereas distribution via the Internet is more difficult to constrain in this way. Other business implications also apply, even down to the particular program aired. Rights management, copyright payments, and control of distribution of copyrighted material are all major factors in what programs and advertisements can be made available to consumers, how much they must pay to view them, or whether it is free of charge. Nevertheless, it is clear that distribution of streaming audio and video media via the Internet is now a major force in program distribution and will need to be accommodated by all broadcasters. It can serve as an alternative distribution medium, provide a value-added service that may increase revenue generated by a given program, and allow distribution beyond traditional borders. See Chapter 6 for technical details on this topic.

STATIONS, GROUPS, AND NETWORKS

The business structure and hierarchy of broadcasting is an important and influential element that must be understood to fully comprehend the industry's operation. It can also vary dramatically between regions and countries of the world.

Terrestrial Broadcasting

Most large towns in the United States have at least one or two local AM or FM radio stations and one or more television stations. Large cities usually have many more. Some of these stations are individually owned, but many belong to station *groups* that also own other stations—in some cases, hundreds of stations nationwide, and in the case of radio, even several within the same city.

Most U.S. TV and radio stations are affiliated with a broadcast "network" (using an older definition of the term, and not to be confused with today's more common computer network terminology). These networks are centralized distribution services that provide content for their affiliated TV and radio stations. Some of these networks own and operate local stations, as well, and these stations are therefore known as "O&Os" (for "owned and operated" stations). In some cases, a local station may be affiliated with more than one network. This is common in radio, but only occurs in the smallest television markets, where there are not enough stations on the air to carry all networks. The number of stations a single entity can own or operate, and the nature of network affiliations, are governed by rules administered by the U.S. Federal Communications Commission (FCC). See Chapter 20 for more detail.

Most U.S. broadcast networks today are associated with a Hollywood content-production studio, and many are also affiliated with a cable content-producing component and/or MSO. Thus, the business of U.S. broadcasting (including many local stations) is dominated by a small number of highly consolidated, multi-tiered businesses.

Most U.S. radio and TV stations are licensed by the FCC as "commercial" operations, in that licensees can support their operations by selling of short-form announcements interstitial to long-form content. Other stations in the U.S. are licensed as "non-commercial" services, which prohibits them from airing regular commercials, although some selling of airtime under more restrictive FCC "underwriting" rules is permitted. The latter stations include religious broadcasters, community-based broadcasters, and "public" radio and TV stations. Because all of those stations are licensed to non-profit entities, many solicit their audiences directly for charitable contributions. All of these business models allow terrestrial broadcasting to be provided free of charge to U.S. audiences.

Local radio and TV stations may also transmit *syndicated programs* obtained directly from program producers or distributors without going through any network. Many stations also produce their own programming, especially local news, traffic information, and weather. Commercial stations sell their own advertising time, with locally originated advertisements transmitted in addition to network- or syndicator-originated advertising. Non-commercial stations are often members of content networks or syndication services that are tailored for non-commercial operations, which offer programs, underwriting announcements, and other opportunities for monetization of content that conform to FCC rules governing such facilities.

Some station-group owners may produce or originate programming (particularly news-type programs) at a central location and distribute it to their individual stations at remote locations. This arrangement is known as *centralcasting*.

Cable and Satellite Networks

The term network is also often used to describe companies that produce one or more program channels for distribution to multiple cable, satellite, and telco operators, but not to terrestrial broadcast stations. Examples of cable and satellite networks are CNN (Cable News Network), ESPN (Entertainment and Sports Programming Network), and "premium" networks such as HBO (Home Box Office) and Showtime; the latter do not include commercials but instead operate by a subscription model. There are many other cable networks of both types.

CHAPTER 3

Sound and Vision

This chapter describes some of the scientific principles that are fundamental to all types of broadcasting. It covers the physical properties of light and sound, and how the basic workings of human hearing and vision allow us to perceive sound from audio signals and moving color images from video signals. We will build on these principles later in the book when we discuss how studio and transmission equipment and systems work.

SOUND, AUDIO, AND HEARING

Sound Waves

As you probably already know, the sounds that we hear are actually pressure waves in the air, which cause our eardrums to vibrate. Everything that produces a sound, whether a guitar string, a jet airplane, a human voice, or a loudspeaker, does so by causing a vibration that sets off the pressure wave.

Sound waves are an example of an *analog* signal—they have continuous and generally smooth variations—and the loudness and pitch of the sound that we hear are directly proportional to the variations in pressure in the air. The *amplitude* of the pressure wave determines how loud it sounds, and the *frequency* determines whether it sounds high or low in pitch. Frequency is measured in *cycles per second*, and this unit is usually known as a *hertz* or Hz. The highest frequency that most people can hear is between 15 and 20 thousand hertz (kHz), although the high range is reduced for most people as they get older. Frequencies below about 30 Hz are felt rather than heard.

A pure tone has a single frequency, known as a *sine wave*. Figure 3.1 illustrates this graphically, as a plot of sound pressure level against time. Typical natural sound is more complex, and is made up from many individual sine waves of different frequencies and amplitudes, all added together. The mixture of different sine waves determines the shape of the wave and the character of the sound we hear.

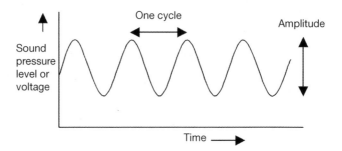

FIG. 3.1.   Analog Audio Sine Wave

Audio

A microphone turns a sound pressure wave in the air into an electrical *audio* signal that matches the amplitude and frequency of the sound wave. In the reverse process, a loudspeaker receives the electrical audio signal and turns it back into sound waves in the air that we can hear. Therefore, Figure 3.1 illustrates both the electrical audio *waveform* for a sine wave, as well as the pressure wave. This electrical audio signal is a varying *voltage* (the unit of "pressure" or amplitude for electricity) that can be recorded and processed in many ways, and ultimately transmitted over-the-air—as discussed later in the book.

The range of frequencies in an audio signal, ranging from the lowest to the highest frequency included, is known as the *bandwidth* of the signal. The range of frequencies that can be passed, combined with information on how accurately their amplitudes are reproduced, is known as the *frequency response* of the equipment or system.

With analog audio, the electrical signal is a direct equivalent of the sound pressure wave that it represents. It is always possible, however, for an analog audio signal to lose some of its frequencies or to have *distortion* and *noise* added to it as it passes through the *signal chain* from source to listener. Losing high or low frequencies makes the sound lacking in treble or bass, respectively; distortion makes the sound harsh and unpleasant, and noise of various sorts makes it more difficult to hear the original sound clearly. Therefore, analog audio is often converted to a more robust digital audio signal for processing and transmission, with many advantages and immunity to those degradations, as described later.

Mono, Stereo, and Surround Sound

Sounds picked up with a single microphone can be carried through a signal chain as a single channel, and end up being played to a listener on a single loudspeaker. Such a source and system is known as *monophonic* or *mono*. The drawback of mono audio is that it does not provide the listener with any real sense of direction or space for the sounds. If the sounds are picked up with two or more microphones, carried to the listener over two separate channels, and played over two loudspeakers, left and right, then it is possible to provide a good impression of the position of the sound in the original studio. Such a system is known as *stereophonic* or *stereo*, and is widely used in audio recording, radio, and television.

Stereo systems can define only the points along a single line between two speakers, however, and are unable to give much of an impression of the acoustics of a sound space beyond that line. A more realistic reproduction of such a space can be achieved by using multiple, properly positioned microphones, a multichannel signal chain, and multiple loudspeakers positioned in front of and around the listener. This is typically called a *surround sound* system.

Today's surround sound systems use four or more channels, designed to feed speakers arranged in front, alongside and behind the listener, all at approximately the same height. Such a system allows sounds to be located anywhere within a single plane around the listener, and represents the state of the art in broadcast sound systems at this writing. Proposed future sound systems may include arrangements with multiple planes of reproduction at varying heights, below and above the listener's head position. These are referred to as "3D" or "Immersive" sound systems, and may involve 10 or more audio channels and speakers placed at appropriate locations in front of, behind, below and above the listener.

LIGHT, VIDEO, AND VISION

Light and Color

Just as sound is a vibration generated by a sound source, which produces a pressure wave that makes the eardrum vibrate, and thus we hear the sound, so too is light perceived by the human viewer in a similar fashion. Light also exists as a type of wave (an *electromagnetic wave*, in this case) that is generated by a light source passing through the lens of the human eye that focuses it on the surface of the eye's retina, which in turn stimulates the various receptors there, creating a sensation that we perceive as sight. Just as the amplitude of the sound wave determines the loudness of the sound, the amplitude of the light wave determines the perceived brightness of the image we see, which in video terms is referred to as the *luminance* level. And just as the frequency of a sound wave determines whether we hear a low or a high tone, the frequency of the light wave determines what color we see, referred to as its *hue*. One other characteristic of colored light is its *saturation*. Saturation refers to the intensity of the color, or to put it another way, how much it has been diluted with white light. For example, a deep, bright red is very saturated, whereas a light pink may have the same hue, but with a lower saturation.

What we see as white light is, in fact, made up of a mixture of many colors that the brain interprets as being white. This can be demonstrated by shining a white light through a glass prism, which splits up the different colors so they can be seen individually. That also happens when we see a rainbow. In that case, each raindrop acts as a tiny prism, and the white sunlight is split up into all the colors of the rainbow. What is perhaps more surprising is that the same sensation of white light in the eye can be produced by mixing just three colors in the right proportion. For light mixing, these *primary colors* are red, green, and blue, also referred to as R, G, and B.

By mixing two of these colors together, we can produce other *secondary colors*, so:

Red + Green = Yellow
Red + Blue = Magenta
Green + Blue = Cyan

By mixing the primary colors in different proportions, we can actually produce most other visible colors. This important property means that the light from any color scene can be split up, using *color filters* or a special sort of prism, into just the three primary colors of red, green, and blue, for converting into video images.

Characteristics of the Human Eye

It was mentioned previously that the retina in the eye is stimulated by light, and we then see the scene. It was also noted that the human eye and brain perceive different light frequencies as different colors. Several other characteristics of human vision are also significant in this process. One of these involves the fact that we see much less detail in our color perception compared to the detail we see in the brightness, or luminance, of a scene. Acknowledgement of this has greatly influenced the way that color signals are carried in color television systems (see the sections on chrominance in Chapter 4 and color subsampling in Chapter 5). Another critical characteristic involves a phenomenon known as *persistence of vision*. After an image being viewed has disappeared, the eye still sees the image for a fraction of a second (just how long depends on the brightness of the image). This allows a series of still pictures in both television and cinematography to create the illusion of a continuous moving picture.

Video

Most people have some idea of what television *cameras* and television *receivers* or *picture monitors* are used for. In summary, a color camera turns the light that it receives into an electrical video signal for each of the three primary red, green, and blue colors. In the reverse process, a television display or picture monitor receives the electrical video signal and turns it back into red, green, and blue light that we can see. Our eyes combine the red, green, and blue images together so we see the full color scene. The video signal that carries the image is a varying voltage, and that signal can be recorded and processed in many ways, and transmitted over-the-air—as described in more detail later in this book.

With analog video, the electrical video signal is a direct equivalent of the luminance of the light that it represents (and in a less direct way, also of the color hue and saturation). As with audio, however, it is possible for an analog video signal to have distortion and noise added to it as it passes through the signal chain from source to viewer. Different types of distortion change the picture in many ways—making it soft and fuzzy, adding a "ghost" image, or changing the colors, for example. Video noise may be seen as "snow" or random spots or patterns on the screen. Therefore, analog video is often converted to a more robust digital video signal for processing and transmission, with many advantages and immunity to those degradations, as described in later chapters.

BASEBAND

The audio and video signals we have mentioned, directly representing sound and image information, are referred to as *baseband* signals. They can be carried as varying voltages over wires and cables and can be processed by various types of equipment. They cannot, however, be transmitted over-the-air by themselves. For that, they need to be combined with radio frequency signals in a special way, called *modulation*, as discussed in the next chapter.

CHAPTER 4

Analog Color Television

NTSC

NTSC refers to the National Television System Committee—the committee that designed the standard for today's analog television system, originally adopted in 1941 (monochrome) and 1953 (color). This chapter explains the principles of television analog video (picture) signals at *baseband* (not modulated onto a carrier). The system for transmitting NTSC signals over the air is explained in Chapter 15.

Frames

NTSC video signals produce a rapid-fire series of still pictures that are shown on a television receiver or monitor. Each of these pictures is called a *frame*. Because of the phenomenon known as persistence of vision, the eye does not see a series of still pictures, but rather an illusion that the picture on the TV is moving.

Figure 4.1 provides a simple example of a series of still pictures that might be used to create the illusion that a ball is bouncing across the screen.

The number of complete pictures presented to the viewer every second is known as the *refresh rate*. If the refresh rate is too low, then any motion on the screen appears to be jerky because the frames do not appear fast enough to maintain persistence of vision; this is known as *judder*. When cinematography was developed, it was found that a rate of 24 frames per second (fps) was enough for reasonably smooth motion portrayal, and that rate has been used for film all over the world ever since. Later, when television was developed in

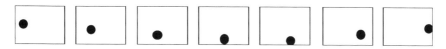

FIG. 4.1. Series of Still Pictures That Create Illusion of Motion

the United Kingdom, a rate of 25 fps was selected, whereas a rate of 30 fps was selected in the United States.

The television picture defined in the NTSC system has a width to height ratio of 4:3; this is known as the *aspect ratio* (see Aspect Ratio section in Chapter 5).

Scanning

In the NTSC system, each video picture frame is "painted" on the television screen, from top to bottom, one horizontal line at a time. This process is called *scanning*. In NTSC, there are 525 horizontal lines in each frame, 483 of which form the actual picture. (The other lines are reserved for auxiliary uses, as described later in this chapter.) The refresh rate is 30 fps.

Interlacing

When the NTSC standard was developed, 30 frames of video per second was about the best that available technology could process and transmit in a 6 MHz-wide television *transmission channel*. Pictures presented at 30 frames per second appear to flicker, however. NTSC television reduces this flicker without actually increasing the frame rate by using an arrangement called *interlacing*. This process divides the 525 horizontal lines in each frame into two groups called *fields*. The first field of each frame presents only the odd-numbered lines of the frame (lines 1, 3, 5...), and the second field presents the even-numbered lines (lines 2, 4, 6...). The two 262.5-line fields interlace together to form a complete image, which is presented as an NTSC frame as illustrated in Figure 4.2. Thus, the NTSC frame rate remains at 30 fps, but the field rate is 60 fields per second, and flicker is thereby reduced in the reproduced video image.

In an interlaced NTSC picture, each of the lines in the picture is still refreshed only 30 times every second. Yet the human eye cannot perceive the fact that two adjacent lines on the video screen are being refreshed at different times when there is only a period of 1/60th of a second between them. The effect is to create the appearance that the full screen is being refreshed twice as often, or 60 times per second, increasing the apparent refresh rate, and greatly reducing the perceived screen flicker.

Film Projection

The concept for such visual "trickery" originated in the cinema world, where an even slower (film) frame rate of 24 fps was used for projection onto a screen. Film projectors use a rotating *shutter* to interrupt the light beam while each film frame is moved into position for projection. The shutter is normally fitted with two blades rather than one, however, thus interrupting the light twice for each projected frame. This produces an apparent refresh rate of 48 times per second,

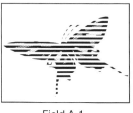 

Field A-1   Field A-2

Frame A

FIG. 4.2.   Two Interlaced Fields Combine to Create Each NTSC
Frame (Line Size Exaggerated for Clarity)

which greatly reduces the perceived screen flicker. A three-bladed shutter is
sometimes used, giving a refresh rate of 72 times per second. Although the cinema industry is also converting to digital projection at this writing, in facilities
where film projection is still employed, the rotating shutter technique remains
in use today.

Progressive Scan

If interlacing is not used with a television picture, then each frame is painted on
the screen, from top to bottom, in its entirety. This is called *progressive scan* and
is not used for NTSC. Progressive scan is used with video for computer monitors and for some digital television formats. In that case, much higher frame
rates are needed if flicker is to be avoided. This also improves motion portrayal.
Fortunately, with computer systems it is actually less complex to implement
higher frame-rate, non-interlaced pictures than it is to implement interlacing.
Progressive scanning can also make the digital compression of video signals
easier (more on this later in this book).

Active Video and Blanking

As previously mentioned, there are 525 horizontal lines of information in an
NTSC video signal, 483 of which are the *active video* lines that carry the actual picture information. The other lines make up the *vertical blanking interval*. Even the

active video lines do not carry picture information for their entire duration, but have a period at the beginning and end of the line known as the *horizontal blanking interval*. The reason for these blanking intervals relates to the way traditional *cathode ray tube* (CRT) television displays, and older-technology television cameras with tube-type light sensors worked before the days of modern flat-panel picture monitors and solid-state cameras.

With CRT displays, an *electron gun* inside the CRT is the device that actually paints the video picture. It shoots a beam of *electrons* (the smallest unit of electricity) inside the tube onto the back of the screen. This causes little dots of chemicals on the screen, called *phosphors*, to generate light. Variations in the strength of the electron beam cause the phosphor to produce different levels of brightness, corresponding to the content of the original scene.

As shown in Figure 4.3, the beam of electrons from the electron gun is a single stream that sequentially illuminates each phosphor, one at a time, across each horizontal line of the screen. The beam moves in a left-to-right, top-to-bottom manner. Each time the beam reaches the right edge of a line on the picture screen, it must stop and then move back to the left-hand side of the screen in order to start painting the next line. If the electron gun were to remain on during this entire process, it would paint a line of the picture on the screen and then immediately paint a streak right below it while it *retraced* its path back to the left side of the screen. In order to prevent this from happening, the electron gun is temporarily turned off after it reaches the far right end of a line, and it remains off until the beam is positioned back on the left side of the screen and is ready to begin painting the next line. The period when the electron gun is off, while it is retracing its route across the screen, is the horizontal blanking interval. It is a very short period, significantly less than it takes to paint one horizontal line of video on the screen.

The same concern about putting a streak on the screen must be addressed when the electron beam finishes painting the bottom line of a video field and needs to be repositioned to the top left-hand corner of the screen to start a new

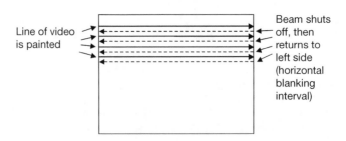

FIG. 4.3. Horizontal Line Scan and Blanking for Four Lines of Video

field. During this operation, the beam is once again turned off while it is retargeted toward the upper left corner, as shown in Figure 4.4. This period when the beam is turned off is the vertical blanking interval or VBI.

The VBI lasts as long as it would take to paint 21 lines on the screen. (Note that these 21 VBI lines are transmitted before the beginning of each *field* of video information, so there are 42 in total per NTSC frame, and hence the 483 active video lines in the 525-line NTSC signal.) These VBI line periods are referred to by their line numbers but contain no picture information. Instead, they are used to transmit other non-video data, the specifics of which have evolved over the period of NTSC's usage. Closed Captioning was added to NTSC transmissions in the late 1970s using the VBI, for example. See Chapter 15 for more detail on usage of the NTSC VBI.

As noted earlier, blanking was deemed necessary when the NTSC video format was developed due to the limitations of tube-type picture displays and television cameras that used tubes as their light-sensing (*pickup*) devices. Most modern cameras use pickup devices like *charge-coupled devices* (CCDs), while modern picture screens typically use *liquid crystal display* (LCD) or *plasma* technology, none of which require blanking intervals for beam retrace. However, to maintain compatibility with all existing devices, the NTSC video blanking timing remains unchanged, and, in any case, some of the NTSC VBI lines are also used for other purposes, as mentioned above. Note that digital television does not include any blanking intervals in the transmitted signal because the picture encoding system is completely different from NTSC, and both video camera and display technologies have moved away from reliance on tube-based systems.

Synchronizing Pulses

Because video relies on a series of sequential images presented in a regular fashion, it is important to keep all components of transmission and reception equipment synchronized in time with one other. Lack of synchronization can cause

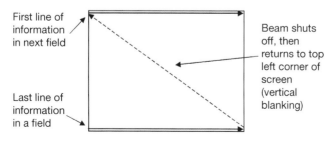

FIG. 4.4.   Vertical Blanking

video frames to be displayed incorrectly, resulting in problems like video "roll-ing," "tilting," or other image distortions. To allow all NTSC video components to be synchronized, both internally (i.e., within itself) and externally (i.e., with other equipment in the production or transmission chain), it is necessary to indi-cate the start of each video frame and line. This is done with *synchronizing pulses*, usually known as *sync pulses*. A *horizontal sync pulse* marks the beginning of every video line, and a series of broad and narrow pulses extending over several lines marks the start of each new video field.

Video Waveform

Let's now look at the actual video signal that carries the picture information, starting with the camera at the beginning of the chain. Typically, a television (video) camera is focused on a scene, and the image is then dissected into scan lines and repetitively into fields and frames.

As shown in Figure 4.5, the camera produces a video signal with a varying voltage on its output. The voltage is proportional to the amount of light at each point in the picture, with the lowest voltage being equivalent to black and the highest voltage being equivalent to the brightest parts of the picture. At the beginning of each line and field, the camera adds horizontal and vertical sync pulses (respectively), so that all equipment using that signal will know where the video lines and fields begin. Ultimately, when this video signal containing pic-ture information and sync pulses is fed to a video monitor, it produces pictures. When it is displayed on a special piece of test equipment called a *waveform moni-tor* that shows the characteristics of the electrical video signal, it is known as a *video waveform*, as shown on the lower right of Figure 4.5.

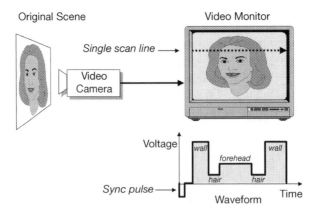

FIG. 4.5. Monochrome Scan-Line Waveform

Luminance, Chrominance, and Composite Video

It is important to remember that—like still photography—the earliest television systems (including NTSC) produced only monochrome images. Color television evolved later, after certain techniques and technologies for monochrome TV had already been established. So it was advantageous to leverage the existing infrastructure by retrofitting color TV systems into the monochrome world. This was one of the key design elements of the NTSC system when it was revised from its original monochrome to its later color format.

For example, the explanation of the video waveform above applies only if the camera is producing a black-and-white image based on the brightness, or *luminance*, of the scene. Cameras producing color images use similar scanning techniques, but generate three separate signals—one each for the three primary colors of red, green, and blue. Given that monochrome TV only needed a single signal path to carry its video, the NTSC standard maintained this tradition by simplifying the routing and transmission of the three R, G, and B signals into one combined electrical signal. This was done by adding R, G, and B in the correct proportions to make the luminance signal, which is always referred to as Y. Then, by subtracting the luminance from each of the red and blue signals, two *color difference* signals are produced, called R-Y and B-Y. These two signals are called *chrominance*.

Again in the interest of leveraging existing infrastructure, and because the eye is less sensitive to color detail than to brightness, the *bandwidth* of the chrominance signals was reduced to make it possible to transmit color in the same transmission channel as monochrome TV. The chrominance signals are then *modulated* (see Chapter 7) onto a *chrominance subcarrier*, also known as a *color subcarrier*. The use of a subcarrier for the chrominance signal allows it to be carried on the same cable with the luminance signal from the camera and around the studio, and ultimately transmitted over-the-air in the same broadcast channel. The resulting electrical signal that combines luminance and chrominance picture information with sync pulses, all carried together on one cable, is known as *composite video*.

Strictly speaking, the electrical video signal components carried through most of the television chain should be referred to as *luma* and *chroma*. These terms have a slightly different technical meaning compared to luminance and chrominance (it has to do with something called *gamma correction*, which is outside the scope of this book). Nevertheless, the terms are frequently used interchangeably—although sometimes incorrectly. For simplicity, we will generally use luminance and chrominance in this book, except in contexts where normal usage is to use the words luma or chroma.

Figure 4.6 shows an example of a composite color waveform carrying the well-known test signal *color bars*. The illustration shows how this signal looks on

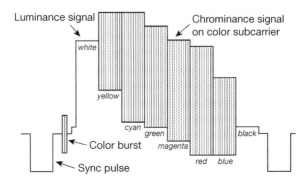

FIG. 4.6.   NTSC Color Bars Waveform

a waveform monitor, while on a picture monitor, all that is seen is vertical bands of color. You can see that each bar has a different level of luminance (brightness), while the chrominance signal, superimposed onto each bar, carries the color information for that bar. The horizontal sync pulse is at the beginning of each line. The short *burst* of color subcarrier that can be seen after the sync pulse falls in the horizontal blanking period, so it is not visible in the video image, but serves to help a color receiver decode the chrominance information.

A special version of this composite video signal, known as *black burst* or *color black* is used in analog television facilities to synchronize different items of equipment together (see Video Timing section in Chapter 9). This signal contains sync pulses and the color subcarrier burst, but no luminance or chrominance signal (hence, it shows as black if displayed on a picture monitor).

Decoding at the Receiver

At the television receiver, after passing through the transmission chain, the composite NTSC signal is decoded. The receiver separates the luminance and color-difference signals, and by adding and subtracting these values the receiver can reproduce the original, separate red, green, and blue signals. These drive the display device to produce the red, green, and blue parts of the picture, which are built up line by line and frame by frame, and which the viewer's eye sees as the original scene. See Chapter 15 for a more detailed explanation of the NTSC transmission signal and the analog TV receiver.

Actual NTSC Frame and Field Rates

The 30 frames per second and 60 fields per second values mentioned previously were defined when the NTSC standard was initially written for black-and-white television. When the color subcarrier was introduced to carry the chrominance information for color television (while maintaining compatibility with existing monochrome systems), it was found necessary to change the frame and field rates slightly—by a factor of just one part per thousand—to help avoid interference with the black-and-white picture and the transmitted audio signal. The actual values used in NTSC systems used today are therefore 29.97 frames per second and 59.94 fields per second. These so-called "non-integer rates" have also been carried forward into ATSC digital TV systems so that content can be converted between analog and digital formats without complex frame-rate conversions being necessary.

PAL AND SECAM

The PAL and SECAM color systems used in many countries outside the U.S. work similarly to NTSC, with the main differences described as follows.

Scanning Formats

NTSC has 525 total lines per picture and 30 frames per second, whereas in most countries PAL and SECAM have 625 total lines and 25 frames per second (525/30 for PAL in Brazil). Note that countries using 25 fps have electrical power supplied at 50 Hz, while N. America and other 30 fps countries have power supplied at 60 Hz. This simplifies the design of equipment used in those countries, but it also reduces problems when shooting video indoors, given that video equipment uses a frame rate that is directly related to the frequency at which light bulbs are flickering.

PAL

The PAL color subcarrier chroma signal has a different frequency from NTSC and reverses *phase* on alternate lines. In the early days of color television, this resulted in less chroma error and more accurate color pictures. With much more stable modern NTSC equipment, this advantage has largely disappeared. There are also minor differences in the way the color difference signals are derived and processed.

SECAM

In SECAM, the color difference signals, instead of being transmitted together, are transmitted on sequential lines, and a memory is used to make them available at the time required. In addition, the color subcarrier is modulated with FM rather than QAM.

Because of difficulties with processing SECAM signals, countries that use SECAM for transmission almost always produce their programs in the PAL standard and convert to SECAM just before transmission.

Channel Bandwidth

NTSC is transmitted in a 6 MHz-wide television channel, whereas PAL and SECAM are usually transmitted in a 7 MHz or 8 MHz channel, depending on the country.

HD ANALOG VIDEO

High definition video, usually known as HD, is generally associated with digital television. Yet some HD-capable video cameras and picture displays (such as the traditional CRT picture tube) operate as analog devices when the images are first acquired, and are finally displayed. This is because the light from the original scene, and the light produced by the final display device, is itself analog. So in some parts of a digital television system (even an HD system), analog video signals may still exist.

Analog high definition video formats are usually referred to with the same terms used for the digital HD formats (1080i and 720p, where "i" means *interlaced* and "p" means *progressive*—more on this in Chapter 5). HD analog connections are sometimes used to feed picture monitors and video projectors, or for other purposes, and may be carried as follows:

- RGB signals: one cable each for red, green, and blue signals
- YPbPr component signals: one cable for the luminance, Y, and one cable for each of the color difference signals known as Pb and Pr

With RGB, the sync pulses may be carried on the green channel, or they may be carried on one or two additional cables (for horizontal and vertical sync). With YPbPr components, the sync pulses are always carried on the Y channel. HD signals use *trilevel sync*, with three levels, rather than the usual two, showing where the different parts of the signal start and finish.

The RGB or YPbPr signals are referred to as *component video* because the three colors, or the luminance and color difference signals, are kept as separate components. This is an important concept, both for analog and digital video. High definition video is always carried in component form; there is no such thing as "high definition composite video."

CHAPTER 5

Digital Audio and Video

This chapter discusses the basics of baseband digital audio and video as typically used in studio facilities. Using digital rather than analog signals in the studio results in higher audio and video quality, greater production capabilities, and more efficient operations no matter whether they feed analog or digital transmission systems.

It should be noted that the terms *digital radio* and *digital television* are usually used to mean the method of transmission of the audio and video signals. It is, in fact, possible for such digital transmitters to be fed with signals produced in an analog studio facility and converted to digital for transmission. For maximum quality, however, it is desirable that both studio production and broadcast transmission be digital.

DIGITAL AUDIO

When we speak of *digital audio*, we are referring to audio signals that have been converted into a series of *binary values* (represented numerically by the digits 0 and 1; see Chapter 6 for more on this), rather than being sent as a continuously variable *analog waveform*. The advantage of this scheme is that binary values can be represented quite simply by either the presence or the absence of a given signal, which is easily detected, and can be distributed and recorded with great accuracy. It is largely a quantitative and not a qualitative process. Digital data also can be easily processed by computers and other digital equipments, making it quite cost-effective given the quantities of scale in today's vast computing marketplace. Digital audio is typically not subject to the usual degradations of analog audio (for example, noise and distortion)—even after being copied many times, and going through a long chain of equipment—although other degradation due to faults in digital processing remain possible.

A useful metaphor often presented to describe the difference between analog and digital compares a bucket of water with a bucket of ball bearings. Determining the exact amount of water that is in the pail is elusive, since some will evaporate, and a little will stick to the sides of the pail when poured out. Every attempt at measurement will likely produce a slightly different result. On the other hand, it is possible to know exactly how many ball bearings are in the bucket, and the number will be the same every time you count them. Thus, ball bearings illustrate the precision and repeatability of digital representation, whereas the vagaries and inconsistencies of analog are represented by the water.

It is also important to recognize that despite the values of digital systems for retention of high quality throughout the many stages of storage, manipulation and transmission of media content, the original input of sound and light information is in analog form, and the end result must also be ultimately reproduced to audiences as an analog presentation. Therefore, the overall quality of the content carried by digital systems relies heavily on the fidelity of conversion to and from analog at the beginning and end of the digital signal chain or process. This procedure is handled by the aptly named *analog-to-digital converter* and *digital-to-analog converter*, which are key components in any digital media operation, as will be described later.

The process of conversion from analog to digital involves five basic concepts that one needs to grasp in order to have an understanding of digital media: *sampling, quantizing, resolution, bitstream*, and *bit rate*. The following section describes these concepts as they apply to audio signals.

Sampling

Figure 5.1 shows an analog waveform plot of the amplitude of an audio signal as it varies in time. In an analog-to-digital converter (ADC), the amplitude of the wave is measured at regular intervals: this is called sampling. Each individual sample represents the level of the audio signal at a particular instant in time.

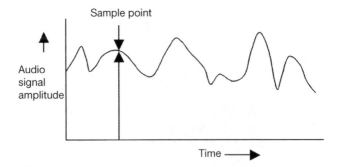

FIG. 5.1. A Single Digital Sample of an Analog Audio Signal

The *sampling rate* is the measure of how often digital samples are taken from the original analog waveform. The more often the original material is sampled, the more accurately the digital signal will represent the original material when it is returned to the analog domain and an analog waveform is generated from the sampled data by a digital-to-analog converter (DAC).

Figure 5.2 shows an analog signal being sampled at some regular interval. Figure 5.3 shows the same analog signal being sampled twice as often. When converting the digital samples back to analog, the DAC will generate an analog waveform that essentially "connects the dots" by taking the shortest path between samples. Note that Figure 5.2 shows several places along the waveform where the original analog signal takes a path that would not be recreated by the DAC because the original waveform does not follow a straight line between samples. The higher sampling frequency in Figure 5.3 captures these variations, however, and a path tracing a straight line between the samples taken in Figure 5.3 will more closely approximate the original analog waveform.

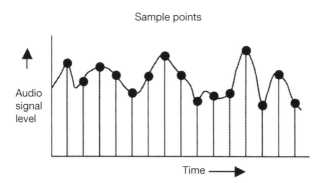

FIG. 5.2. Periodic Digital Samples of an Analog Signal

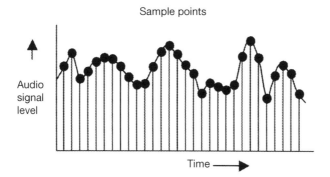

FIG. 5.3. More Frequent Digital Samples of the Same Analog Signal

Five different standard sampling rates are typically used for digital audio in studios: 32,000 samples per second, 44,100 samples per second, 48,000 samples per second, 96,000 samples per second, and 192,000 samples per second. Usually these rates are referred to simply as 32 kHz, 44.1 kHz, 48 kHz, 96 kHz, and 192 kHz, respectively. Audio compact discs (CDs) use a sampling rate of 44.1 kHz, while most broadcast digital audio studio equipment uses 48 kHz sampling or higher rates.

Quantizing

The audio samples are converted to a series of discrete values, represented by binary numbers ranging from 0 to the largest binary number used by a digital audio system. For each sample, the value nearest in size to the actual analog amplitude is used. This is known as quantizing.

Resolution

The resolution of digital audio is the precision with which the amplitude of sampled audio is measured. In other words, the resolution used by the digital audio system defines how accurately the digital audio sample values represent the original analog material. Like many aspects of digital systems, resolution is measured in *bits*, because they are the units of which binary numbers are composed. The higher the number of bits available in the system, the more accurately the digital signal represents the original analog material. The audio CD format (often referred to as *Red Book*, after the document that originally specified the format in 1980, and also which is now employed by many other professional audio hardware and storage systems besides those associated with CDs) uses 16-bit binary values to represent each audio sample. This means that each of the 44,100 samples taken from every second of an audio waveform has its amplitude value represented by a binary number that is 16 digits long. The resolution of such a system is thereby defined to have 2^{16} possible values, or 65,536 discrete amplitude steps in decimal terms. Thus, each audio sample is assigned the closest amplitude value that the system can determine from the 65,536 possible values available. Given the tiny variations between each of those many possible values, it is likely that any amplitude encountered by the system would be able to be assigned a digital value closely representative of its exact actual analog level, and thus such fine-grained resolution produces very high-quality audio.

Note that 16-bit audio is not just twice as accurate as 8-bit audio in replicating the original analog material—it is 256 times better. Chapter 6 explains that adding one more bit to each sample number <u>doubles</u> the accuracy of that number; therefore, a 9 bit sample is twice as accurate as 8 bits, and 10 bits is twice as accurate as 9 bits. Continuing in this way, it can be seen that 16-bit resolution is 256 times more accurate than 8 bit resolution.

Audio Bitstream and Bit Rate

When binary numbers representing the audio samples are sent down a wire one after the other, the stream of binary digits (bits) is referred to as a *serial bitstream*, or more commonly, simply a bitstream.

The bit rate necessary to transport a digital audio signal is directly related to the digital resolution of the digital audio and its sampling rate. Using the digital resolution and the sampling rate for CDs, for example, we can calculate the bit rate necessary to transport CD audio:

> *CD digital resolution: 16 bits/sample*
> × *CD sampling rate: 44,100 samples/second*
> = *CD bit rate per channel: 705,600 bits/second/channel*
> × *2 stereo channels: 2*
> = Total CD bit rate: 1,411,200 bits per second or about 1.4 Megabits per second

Note that bit rates are important when specifying the bandwidths required by digital transmission paths, and they are measured in bits per second. Meanwhile, the storage of digital signals (on magnetic hard-disks, optical disks, solid-state memory chips, etc.) is measured in the next-highest binary unit, the *byte*. There are eight bits in each *byte* of data. So, in order to store one second of CD stereo audio on a disk, we need 1,411,200 ÷ 8 = 176,400 bytes of disk space. A typical three-minute song, therefore, requires 176,400 bytes × 180 seconds = 31.752 Megabytes (MB) of disk space. (A good rule of thumb for storage of audio in the CD format is 10 MB per minute.) Various *data compression* techniques have been developed for reducing the bit rate and size of audio files, and these are covered later in the book. For more on bits and bytes, see Chapter 6.

Analog/Digital Conversion for Audio

As noted earlier, the process described above is carried out in ADC. The process can be reversed in a DAC, which then re-creates an analog audio waveform from the digital bitstream. These converters may be stand-alone units or may be inside other equipment.

It is common for equipment that uses digital signals internally, such as a portable digital audio recorder, to have an ADC and DAC inside the unit. This means that the external interfaces, or connections to the outside world, can be either digital or analog. This gives much greater flexibility in making system interconnections to the device. As an example, a consumer CD player, which plays back digital audio from the disc, usually connects to a stereo system through analog left and right channel connections. The same player may also have a digital *coaxial* or *optical* audio output, however, which can feed a home theater system with higher-quality digital signals, thereby avoiding multiple conversions from digital to analog and back, which can have negative effects on overall signal quality. Similarly, professional digital audio equipment can usually work either in all-digital systems or as part of a mixed analog/digital environment.

AES3 Digital Audio Distribution Standard

The *AES/EBU* digital audio format, more commonly known today as *AES3*, is a standardized format for transporting digital audio from place to place and is the most common standard used for this purpose. AES/EBU refers to the Audio Engineering Society and the European Broadcasting Union organizations, respectively, which together developed and first published the standard in 1985. (It has undergone several subsequent revisions.)

In order to transport digital audio information, a stream of digital bits must be carried from the originating point to the receiving point. So that the device receiving the bits can understand which ones belong where, a standardized format for transporting the bits must be defined. This is what AES3 does. The format is able to carry two channels of audio (either two mono channels or one stereo pair), at any of the standard digital audio sampling frequencies, and with an accuracy of 16, 20, or 24 bits per sample.

The stream of digital bits is organized into 64 bit segments called *frames* (not to be confused with video picture frames). Each of these frames is further broken down into two subframes. Subframe 1 carries the digital audio information for audio channel 1, and subframe 2 carries the digital audio information for audio channel 2. In the vast majority of radio stations broadcasting music, the two subframes correspond to the left and right channels of the stereo audio pair. The AES3 frame structure is illustrated in Figure 5.4. The audio data is sent using a system called *pulse code modulation* (PCM), and this type of uncompressed audio is often known as *linear PCM*.

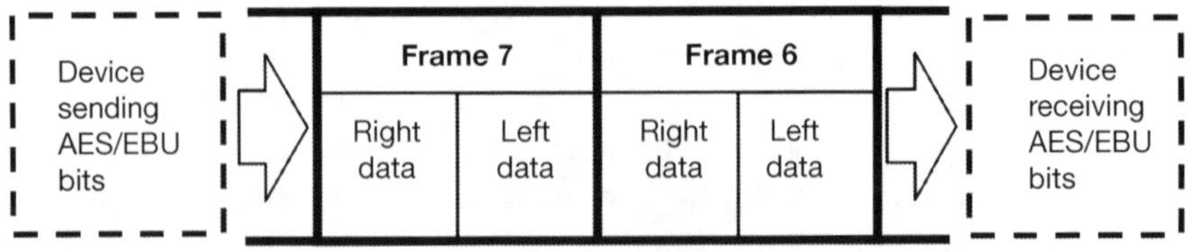

FIG. 5.4.   AES/EBU (or AES3) Digital Audio Frame Structure

AES3 signals can be carried over *twisted pair* or *coaxial* cables, or on *fiber-optic* cables. They may also be combined (embedded) with video signals, and also may be sent on some types of RF-based distribution systems. It is important to understand that the AES3 standard only defines the way digital audio is transported from one point to another; it does not define the format for digital audio storage, which is covered in Chapter 8.

SPDIF

Another digital interface you may come across is the Sony/Philips Digital Interface (SPDIF, or sometimes S/PDIF). It can use either coaxial or fiber-optic cables. Because it is intended for the consumer marketplace, SPDIF is specified to use simpler and less expensive wiring and connectors than the professional AES3 standard, but the data format of signals is quite similar between the two systems. In addition to carrying a linear PCM AES3 signal, SPDIF can alternatively be used to carry a compressed audio bitstream such as *Dolby AC-3*, which may include *5.1 channel surround sound* (see AC-3 audio section in Chapter 16).

Digital Signal Robustness

Digital signals are much more robust than analog, and they retain their quality though multiple recordings and transmission. The reason for this is that digital equipment only has to store and reproduce the binary 0s (low level signals) and 1s (high level signals) to represent the content. It does not need to accurately reproduce all of the levels between low and high as it would for a continuously variable analog signal. To see how this works, let's look at a digital transmission through a noisy channel (this applies generally to both audio and video, so the transmission could be of either type).

Figure 5.5 shows a transmitted string of "0s and 1s" as they might be received, represented by high and low signal levels. So long as the receiver can distinguish between the high and low levels, the signal received will be precisely the same as the series of 0s and 1s that was transmitted. Drawing (a) shows the digital signal with a small amount of noise in the channel, represented by the variable dark line near the bottom of the waveform. Drawing (b) shows the same digital signal with a much larger amount of noise in the channel—at about half the amplitude of the wanted signal. In an analog system, this amount of noise would be unbearable, almost drowning out the wanted signal but, as is clear from the drawing, it is still easy for a digital receiver to detect and distinguish the low- and high-level signals, so the 0s and 1s they represent will be recovered perfectly. Only when the noise level increases greatly, to nearly the same level as the high-level signal (as shown in the lower drawing of Figure 5.5), does it become

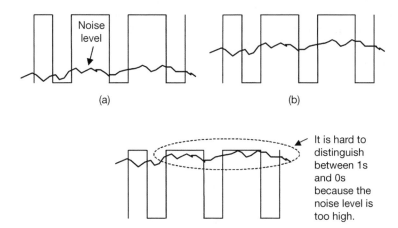

FIG. 5.5.   How a Digital Signal Relates to Noise

impossible to accurately detect the digital signal, because it becomes too hard to determine when the signal level is intended to represent a 0 or a 1. Thus digital signals, whether audio or video, are much less affected than analog signals by any noise introduced in a transmission path or recording device.

SD AND HD DIGITAL VIDEO

Each line in an analog NTSC video picture is a continuous stream of video information. In contrast, each line in a digital video picture can be considered as a series of discrete *pixels*, an abbreviation derived from the words *picture* and *element*. The basic principles of converting baseband analog video to digital video are in many ways similar to the digital audio conversion described previously. The analog video waveform of every line in every frame is sampled at a suitable rate and resolution. As with digital audio, the video samples are thereby converted into a series of binary values; they are then coded in a special way and finally produce a serial digital video bitstream.

Digital Components and Color Subsampling

Modern digital video systems sample the *luminance* and *color difference* video component signals separately and produce *digital component* video signals. This allows much higher picture quality to be maintained compared to NTSC *composite video*. Once in the digital domain, special methods can be used to send the separate components in order, one after the other, down a single wire (a process called *serializing*), thus maintaining the single-wire convenience of composite video.

As with NTSC video, and again owing to our reduced visual acuity to color versus brightness, digital color difference signals can have their bandwidth reduced compared to the luminance signal, to save on transmission bandwidth and storage space. In digital video, this is done by sampling the color difference signals less often than the luminance signals and is known as *color subsampling* or *chroma subsampling*.

There are various methods of such subsampling, and the four most commonly used are shown in Figure 5.6. The nomenclature identifying a given subsampling scheme uses three values separated by colons, which is actually the statement of a ratio of sampling frequencies, simplified into the form $a{:}b{:}c$. The first value, a, is by convention always 4, and it sets a reference for the number of pixels on a given line of video. All of these pixels have a discrete luminance value in the digital signal, but what the values b and c effectively indicate is how many of these pixels also have their color information discretely captured, in the horizontal (b) and vertical (c) dimensions, respectively. Value b indicates the number of pixels that have their color difference signals sampled in the *same* line of video as those four pixels referenced, and value c indicates the number of color samples taken for those four pixels in the *next* line of video (i.e., the four pixels directly under the referenced pixels).

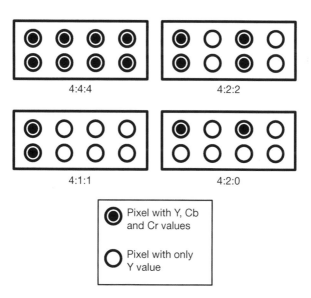

FIG. 5.6. Simplified Depiction of Digital Video Color Subsampling

Thus a 4:4:4 system is not subsampled at all—every pixel has both its luminance and its two color difference samples captured, whereas a 4:2:2 system has its color difference values sampled only on two of the four pixels in each line. The 4:2:0 system also samples color difference on two out of every four pixels, but it only does so on alternating lines (the zero value for c indicates that no color samples are taken for the four pixels directly under the referenced pixels). The 4:1:1 system samples color once for every set of four luminance samples.

Note that this does not mean that there is no color displayed on the viewer's screen for those pixels that are not color-sampled. In fact, the color values are *interpolated* across adjacent pixels. This simply means that a pixel without discrete color information of its own simply repeats the last color value sampled in a previous pixel next to or above it, whereas each pixel on the screen retains its own discrete luminance level. Another way of thinking of this is that in a color-subsampled system, the "color pixels" are bigger than (and a bit offset from) the "monochrome pixels," since certain pixels repeat the color values of their neighboring pixels in one or both dimensions.

Specifically this means that a 4:2:2 system has half the resolution for color than it does for luminance in the horizontal dimension only (but retains full color resolution vertically), while 4:2:0 has half the resolution for color compared to luminance in both horizontal and vertical dimensions. The 4:1:1 system maintains full color resolution vertically but provides only a quarter of the horizontal resolution compared to luminance.

The 4:2:2 method is commonly used for high-quality professional video, and 4:2:0 is typically used for DTV transmission, as well as for the DVD and Blu-ray optical disk formats. Some digital camcorders use the 4:1:1 system.

Digital Video Formats

Analog video signals produced by 525-line cameras, as used for NTSC television, can be sampled and converted into digital video. This process produces one of the *standard definition* DTV formats. But because the DTV transmission standard includes a method for *compressing* the transmitted video, it allows more video information to be squeezed into the same broadcast transmission channel size as was used for analog television. This provides several opportunities for new and improved video formats that were not available with the NTSC standard. Specifically, it allows for higher *resolution* pictures, with wider *aspect ratio* and, in some cases, with higher *frame rates* and/or *progressive scanning*.

The ATSC digital television standard defines 18 different picture formats for transmission (or 36 if all the same formats using a small frame rate adjustment mentioned later are counted). Only a few of these formats are generally used by broadcasters, however.

Like NTSC video signals, all of the digital video formats end up as a rapid-fire series of still pictures that are displayed by a television receiver, but there are significant other differences between NTSC and digital video, as will be illustrated below.

The main characteristics of a digital video format are as follows:

- Number of active lines per picture
- Number of active pixels per line
- Frame rate
- Interlaced or progressive scan
- Picture aspect ratio

Citing each of the parameters in the list above, the video production formats (i.e., those used by cameras and recorders, etc.) most likely to be found in broadcast facilities are the following:

High Definition
- 1080 lines × 1920 pixels, 60 frames per second, progressive, 16:9 aspect ratio (usually referred to as 1080p60)
- 1080 lines × 1920 pixels, 30 frames per second, interlaced, 16:9 aspect ratio (usually referred to as 1080i)
- 1080 lines × 1920 pixels, 24 frames per second, progressive, 16:9 aspect ratio (usually referred to as 1080p24)
- 720 lines × 1280 pixels, 60 frames per second, progressive, 16:9 aspect ratio (usually referred to as 720p)
- 720 lines × 1280 pixels, 24 frames per second, progressive, 16:9 aspect ratio (usually referred to as 720p24)

Standard Definition
- 483 lines × 720 pixels, 60 frames per second, progressive, 16:9 aspect ratio (usually referred to as 480p)
- 483 lines × 720 pixels, 30 frames per second, interlaced, 16:9 aspect ratio (sometimes referred to as 480i widescreen)
- 483 lines × 720 pixels, 30 frames per second, interlaced, 4:3 aspect ratio (sometimes referred to as 480i)

The high definition (HD) formats have 1080 or 720 lines, and the standard definition (SD) formats all have 483 lines. The numbers of lines and pixels for the SD production formats are slightly different from those actually transmitted in ATSC DTV, but this is not significant for most purposes. The last format listed above is the nearest digital equivalent of analog NTSC video and is often referred to as *601 video*—from the ITU (International Telecommunications Union) BT. 601 standard for 525- and 625-line digital video sampling and format.

For various reasons mainly to do with transferring programs into different formats for distribution in different markets, there is an increasing move to produce some types of programs at 24 frames per second—the same as film rate for motion pictures. This frame rate is therefore common in many high definition production and postproduction facilities.

Advanced Formats: Mobile, Internet, and UHDTV

Newer forms of television distribution such Mobile DTV and Internet TV may use video formats with less lines and pixels than standard definition video such as QVGA and QCIF. This is intended to maximize the efficient use of transmission bandwidth and takes into account the fact that these services are often displayed on smaller screens. Such program material is produced as either SD or HD video and converted to the necessary format prior to transmission or distribution. More on these formats can be found in Chapters 16 and 17.

Meanwhile, there are also imaging formats with <u>more</u> lines and pixels than the high-definition formats listed above, with 2,048 pixels on 1080 lines, or 4,096 pixels on 2160 lines (known as "2K" and "4K," respectively) which are used for *digital cinema* production and display. There are also emerging video formats known collectively as *UHDTV* (for ultra-high definition television), one using 2160 lines and 3840 pixels (sometimes referred to as *UHD-1*, or simply "4K," although as can be seen, slightly different from the "4K" used in digital cinema), and another with 4320 lines and 7680 pixels (sometimes referred to as UHD-2, or simply "8K"). At this writing, UHD-1 is becoming available in high-end consumer displays, but content in the format remains quite limited, and broadcasting in this format is not yet possible. For this reason, consumer UHD displays typically include *upconversion*, which presents HD content with artificially increased resolution.

UHD-2 remains an experimental format at this writing, with most of the development coming from the Japanese public service broadcaster NHK, under the name *Super Hi-Vision*. It is intended for display on very large screens and allows close-up viewing to provide an intense impression of reality and immersive visual effect. (The system also includes a 22.2 channel audio system for an equally intense and immersive sonic experience.) In the long term, this format might be used for broadcast purposes, although there are many challenges for recording, distribution, and transmission. In the meantime, it can be used for special purposes such as exposition displays and museum installations.

UHDTV formats may also include other enhancements over HDTV beyond spatial resolution enhancements, such as improved color resolution using 4:2:2 or 4:4:4 subsampling as described earlier, plus faster frame rates and higher sample resolution ("bit depth"), as discussed later. Another improvement could include wider *color gamut* by which a broader range of the human-perceptible colors that exist in nature—but are not possible to show in current DTV formats—might be represented. More on UHDTV can be found in Chapter 18.

3D TV

Another variant of video involves 3D imaging, which can be applied to either SD or HD formats. There are two primary methods of producing images that appear three-dimensional to the television viewer. One, called *stereoscopic*, requires the viewer to don eyewear of some kind, which causes the viewer's two eyes to each see different elements of the content displayed on the screen. This provides the sensation of depth via *parallax*, that is, slightly different viewing angles of objects on the screen presented to each eye, which is how we achieve depth perception in real life.

So-called *passive* stereoscopic 3D eyewear uses some sort of filter (different on each eye) to separate the two images presented simultaneously on the screen into a single image for each eye. Early 3D systems used the *anaglyph* technique (with different colored filters on each eye, used mostly on projection systems in cinemas), wheras more recent passive stereoscopic systems use *polarizing filters* (with different polarization of light sent to each eye, used in most passive 3D TV systems). *Active* stereoscopic 3D TV systems use eyewear with electronic "shutters" that allow alternate frames to be seen by each eye (the left eye's shutter is open while the right eye's shutter is closed, and vice-versa). Active 3D eyewear requires a power source (typically a rechargeable battery). In all cases, the type of eyewear must be matched to the screen's method of display, so 3D televisions are typically sold with several sets of the proper eyewear included. 3D televisions may be used for display of regular ("2D") television content without eyewear. Most 3D televisions also include features for adding "pseudo-3D" imaging to 2D content.

The other primary approach is called *auto-stereoscopic* 3D (sometimes called "auto-3D"), which does not require eyewear. In this case, the screen of the TV display includes a prismatic lens that splits the two images up into small vertical slices, in a fashion similar to the way a greeting card or a graphic in a children's book appears to show two different images depending on the angle at which it is held. Auto-stereoscopic TV display screens are designed so that a viewer's eyes will each fall into adjacent vertical bands to get different images. Optimum viewing typically occurs only within a certain distance and direction from the screen.

While standards for ATSC terrestrial broadcast of 3D television content are still under development at this writing, certain cable or satellite TV channels and Blu-ray discs have offered 3D content. In cable and satellite systems, video formats have been retrofitted to incorporate the two images in *frame-compatible* ways, either by splitting each image frame horizontally (called "top and bottom" [TaB]) or vertically (called "side by side" [SbS]), with each eye-view utilizing one half of the screen—a process referred to as *frame-packing*. This results in a reduction of the spatial resolution of the image by one-half in either the horizontal or the vertical dimension. The Blu-ray 3D format uses an extra-large frame for 3D, so the original image resolution is retained. In all cases, however, the 3D television then displays the two views as full-screen images in alternating frames, thus reducing the frame rate seen by each eye to one-half the original rate.

Viewing 3D content without glasses on a stereoscopic display (with its 3D feature turned on) will result in a double image. At this writing, proposals for future systems have been made that would allow a *service-compatible* approach, by which a single content feed could be viewed in 2D or 3D at the users' option. But today, 2D and 3D versions of a given content item each require separate encoding and transmission (or packaging). Most 3D content can still be viewed in 2D with the 3D feature on the receiver turned off, but any spatial resolution penalty would still apply.

Thus, 3D TV presentation at this writing pays a temporal resolution penalty—and in many cases a spatial resolution penalty, as well—to enable the depth of imagery. For many viewers, this has proven to be an unacceptable tradeoff. Given that auto-stereoscopic displays remain quite rare and expensive at this writing, wearing the glasses required for stereoscopic 3D TV viewing is also an unwelcome burden for many viewers. Some viewers have also reported eye strain, headaches, or motion sickness from watching 3D TV. As a result, at the time of this writing, 3D TV has not received a high acceptance rate among consumers, and some early 3D TV services delivered via cable or satellite TV have been suspended or cancelled.

Let's now look further at various digital video characteristics.

Lines and Pixels

These attributes determine the *resolution* of the picture, which largely affects how much detail it can display (although there are several other factors influencing how sharp the picture looks). As learned earlier, for digital video, the number of lines is the number of active lines in the picture (excluding the vertical blanking interval, if any), and each line is divided up into discrete pixels. A pixel may be considered as a dot (although it is usually square in shape) and actually includes three separate components for red, green, and blue light emissions, which together are all considered to be one pixel—the smallest component of a video image. Pixels have become familiar to most computer users because the resolution of a computer screen (i.e., the level of detail on the screen) is usually defined in terms of pixels. Some typical computer screen resolutions, defined in terms of horizontal and vertical pixels, are 640 × 480 and 1024 × 768. The more pixels there are on the screen, given a fixed monitor size, the more detailed and sharper looking the image will be. There is a trade-off, however: the larger the numbers of lines and pixels per frame, the more data is needed to transmit or store the video image.

The highest definition ATSC video format has 1080 lines on the screen. Each of these lines contains 1920 pixels, so there are about 2 million pixels in total. The standard-definition ATSC video format has 480 lines, each with 720 pixels, so there are about 340,000 pixels in total. This is why it is sometimes said that high-definition television is six times as sharp as standard definition. In fact, when compared to NTSC standard definition video, the difference is even greater because the analog NTSC system in practice does not carry the full resolution of digital SD video.

Frame Rate

There are three standard frame rates for ATSC digital video: 24, 30, and 60 frames per second. Twenty-four frames per second is the rate commonly used for film, and this may be used for video carrying film-originated material or, more recently, for actual video production. Thirty frames per second is the same as the rate used for NTSC video. Sixty frames per second is a faster frame rate that improves motion rendition and reduces screen flicker by refreshing the video screen more often. The disadvantage is that once again, the more frames per second that are transmitted, the more data that is needed to transmit or store the video signal.

All of these frame rates mentioned are used for DTV transmission. The 30 and 24 frames per second rates with progressive scan may, in some cases, be used for production, but they are not used for display at the receiver because, as previously mentioned, the picture would flicker unacceptably. In these cases, the receiver converts these signals to either 30 frames per second interlaced or, in some cases, 60 frames per second progressive, for display to the viewer.

For the time being, at least, most television facilities also use frame rates that are one part per thousand smaller than the integer frame rates listed previously. They use these so-called *fractional frame rates* to facilitate conversion of NTSC pictures having a frame rate of 29.97 frames per second (see Chapter 4) to digital video for ATSC transmission and to avoid the complication of having multiple timing signals within a single facility.

Interlacing

Each of the three frame rates in ATSC video can be employed with either interlaced or progressive scanning, except that the 1920 × 1080 format is not available with 60 frames per second, progressive, and the 1280 × 720 format is available only with progressive scan. Thus, ATSC does not support 1080p or 720i formats.

There has been much debate over whether interlaced or progressive scanning is preferable for digital television. It is generally accepted that 60 frames per second progressive scan video produces the best motion rendition, and this has an advantage for high-motion programming such as sports. In the ATSC standard, however, the maximum resolution format possible with 60 frames per second progressive has 720 lines because, at the time the standard was developed, the bandwidth required for 1080 progressive lines at 60 frames per second was too great to be recorded or transmitted. Under critical viewing conditions, particularly with diagonal lines and/or high motion, some interlacing *artifacts* may be visible with 1080i. In reality, both 1080i (interlaced) and 720p (progressive) formats are capable of producing extremely high-quality pictures, and the choice of whether to use interlaced video or noninterlaced video at a broadcast facility is, to some extent, a question of personal preference.

Aspect Ratio

Two picture aspect ratios, 4:3 and 16:9, as shown in Figure 5.7, are specified in the ATSC standard and are used for digital video.

The 4:3 width-to-height ratio was selected because it is the same as used in NTSC video, and a tremendous number of archived television programs are in this format. The 16:9 ratio was selected for all high-definition programming and, optionally, for standard definition. It was to some extent a compromise between the motion picture industry's desire for as wide a screen as possible and the manufacturing costs of tube-based displays. About 80 percent of motion pictures are shot at an aspect ratio of 1.85:1, which easily fits into a 16:9 screen with negligible use of *letterboxing* (a technique used to fit a video image onto

4:3   16:9
NTSC or ATSC   ATSC

FIG. 5.7.   4:3 and 16:9 Aspect Ratios

a television screen, without altering the aspect ratio of the original video image, by blacking out the top and bottom portions of the frame).

Video Bit Rates

Typical figures for sampling and bit rates of digital video in the studio are as follows:

High Definition Video
    Sampling rate: 74.25 MHz
    Resolution: 10 bits
    Samples per pixel: 2 (average for 4:2:2 sampling)
    Total HD bit rate: $74.25 \times 10 \times 2 = 1.485$ Gigabits per second

Standard Definition Video
    Sampling rate: 13.5 MHz
    Resolution: 10 bits
    Samples per pixel: 2 (average for 4:2:2 sampling)
    Total SD bit rate: $13.5 \times 10 \times 2 = 270$ Megabits per second

Remember that, in both cases, these are the bit rates for raw, uncompressed digital video. By using compression (discussed in Chapter 16), these rates may be greatly reduced for transmission or storage.

Analog/Digital Conversion for Video

As with audio, the video analog-to-digital conversion processes are carried out in ADCs and DACs. As before, these may be stand-alone units or may be inside other items of equipment.

It is possible for equipment that uses digital signals internally, such as a *digital video recorder* (DVR), to have ADCs and DACs combined with other processes inside the unit. This means that the external interfaces, or connections to the outside world, can be either digital or analog. Again for systems that still use analog, this gives much greater flexibility in making system interconnections, which can be either analog or digital. As an example, a standard definition digital VTR, which records and plays back digital video from the tape, may be incorporated into either an analog or digital video distribution system. Modern SD and HD video systems usually use digital distribution throughout, however (and remember that there is no such thing as a single-wire digital composite HD signal).

Every conversion to or from analog introduces a small quality loss. Therefore, for highest quality, the number of conversions backward and forward between analog and digital video should be kept to a minimum.

SMPTE Serial Digital Interfaces

Standards for transporting uncompressed digital video from one place to another in a broadcast studio are set by the Society of Motion Picture and Television Engineers (SMPTE), which has also developed many of the other standards on which television production relies. *SMPTE 259M* specifies the serial digital interface (SDI) for SD video, and *SMPTE 292M* specifies the high-definition serial digital interface (HD-SDI) for HD video. *SMPTE 424M* and *SMPTE 425M* are the standards for a 3 Gb/sec interface that can carry 1080p60 and 2K signals. At this writing, it is expected that similar standards will be developed for transporting UHDTV content.

The HD-SDI or SDI format carries the stream of digital bits representing the luminance and chrominance samples for the video. The analog sync pulses are not digitized, so special codes are added showing the start of active video (SAV) and end of active video (EAV). The luminance and chrominance data is combined together, and finally the stream of digital bits is sent sequentially down the cable. HD-SDI and SDI are similar in the way they work; the main difference is the much higher bit rate needed for HD-SDI.

Using these standards, digital video signals can be carried over coaxial cables or over longer distances using fiber-optic cable. They may also be sent on some types of RF-based distribution systems. It is important to understand that the SDI and HD-SDI standards only define the way uncompressed digital video is transported from one point to another within professional facilities; they do not define the formats for storage on videotape and servers, which are covered in Chapter 9, nor do they define how digital video is transmitted via broadcasting, which is covered in Chapter 16.

AUDIO AND VIDEO DATA COMPRESSION

As explained earlier, digital audio and video signals generate sizeable amounts of data, and this data requires substantial bandwidth for transmission, and large amounts of storage for recording. To reduce these requirements somewhat, much development has taken place on the reduction of digital audio and video data to lower and more manageable rates. This general term for any such reduction in bit rate or file size after initial creation of the bit stream or file is *data compression*. That term is typically applied to such efficiencies applied to digital audio and video, but highly specialized techniques are used in these cases, and they are more properly called digital audio or video *bitrate reduction*, or *perceptual coding*, for reasons that will become clear. (Sometimes the term *source coding* is also used.) Because these systems are applied at the two ends of a transmission or storage process, they are considered as an (en)code/decode process and are thus generally called *codecs* (for <u>co</u>der/<u>dec</u>oders).

With any sort of data compression, two general approaches are possible: *lossless* and *lossy* coding. As the name implies, lossless coding applies data compression in a way that is fully reversible, in that the compression applied to a bit stream or data file can be undone at any time, and the original bit stream or file can be recovered or reconstructed with "bit-for-bit" accuracy—a complete mirror of the original, at its original file size or data rate. Lossy coding, on the other hand, compresses a file or bit stream in ways that cannot be reversed, such that the data removed from the file or stream to make it smaller is lost and never recovered. Lossy coding can produce smaller files and lower bitrates than can lossless coding, but lossy coding may produce noticeable degradation or *artifacts* in the decoded audio or video signals if data lost in the process is excessive.

To maximize the amount of data compression that can be applied without significant artifacts, and thereby improve the efficiency of transmission or storage of digital audio and video content, specialized lossy coding systems have been developed, exclusively tailored for use on digital audio or video signals. These systems can achieve relatively high degrees of compression as a result of designs that exploit certain weaknesses in human perception of sound or images. They ensure that the data removed from the signal degrades only certain elements of the sounds or images that the typical listener or viewer will not notice. As a result, the use of this class of codecs is often referred to as "perceptual coding."

The first perceptual codecs were developed in the late 1980s, and they played a pivotal role in enabling the introduction of digital audio and video broadcasting, along with other forms of digital media content distribution. Improvement of these systems continues through the present day, such that high-quality audio and video signals can be transmitted at ever-decreasing bit rates, or occupying ever-smaller storage space. Without the significant efficiencies these codecs provided, digital radio and television broadcasting as we know it today would not be technically or financially feasible. While lossless coding rarely provides a reduction in bit rate or file size greater than 3:1 (with 2:1 reduction at best being more typical), at this writing perceptual coding routinely reduces bit rate or file size by factors of more than 10:1 for digital audio, and by around 100:1 for digital video, with advanced systems capable of twice those factors, and proposed future systems expected to double even that efficiency. These codecs thereby reduce the studio bit rates associated with HD video from around 1.5 Gbps to under 20 Mbps, and for stereo audio from 1.5 Mbps to under 100 kbps at this writing.

Numerous perceptual codecs have been standardized, most notably by the *Moving Pictures Expert Group* (MPEG), a component of the International Standards Organization (ISO). Other coding *algorithms* have been developed by individual companies and remain proprietary. For broadcasting, the use of standardized coding is essential, since broadcasters only control the encoding side of such a system in their transmissions and rely on consumer electronics manufacturers to build corresponding receivers to perform the decoding.

There is an important caveat to keep in mind regarding perceptual coding, however. First, while a single pass through a perceptual codec may not result in any noticeable degradation to the content, repeated application of (the same or different) perceptual codecs to the same audio or video stream or file can create perceptible artifacts. In some cases, these artifacts can become severe after only a few encode/decode passes. Most encoders can apply varying amounts of compression to signals passing through them, and the less compression applied, the less likely artifacts are to arise. It is, therefore, recommended that as few generations of compression as possible should be used on a given signal path, and in each encoding case, the least amount of compression as is feasible should be applied.

CHAPTER 6

Information Technology

Broadcasting today is heavily dependent on computers and the associated *information technology* (IT). This is partly for the regular office programs used for station management and partly for the specialized software applications that are increasingly used for systems such as program planning, traffic, electronic newsrooms, and station automation. In addition, much of the equipment used for producing and processing digital audio and digital video is based on computer technology. Also, as we learned in the previous chapters, digital audio and digital video signals are defined using the same terms used for computer data. For all of these reasons, it is worth understanding some of the basic terms and components that relate to the IT side of broadcast engineering.

All computer systems and networks, including the Internet, use binary data, that is, data that can be in either of two states; these two states are generally represented by zeroes and ones. The unit of such data is the *bit*, the value of which, accordingly, can only be 0 or 1. Computer-based systems use such bits to process, store, and carry different types of information, whether it is a text document, numerical spreadsheet, database, graphics, audio recording, or moving pictures. All of these types of information can be coded as a series of numbers, which is how they become types of digital data. So let's start with the numbering system used and what we actually mean by bits and *bytes*.

BINARY

Binary Numbers

The regular decimal numbering system that we use every day has 10 digits, 0 to 9, and when counting, we carry (add) 1 to the next column each time we reach the largest digit, which is 9. In electronic computation and its associated circuitry, it's much easier if we only have to deal with two states—off/on or low/high—rather than ten states, so, as mentioned earlier, most digital systems use a numbering system called *binary* that only needs two digits: 0 and 1. When counting in binary, we still add 1 to the next column each time we reach the largest digit, in this case 1. Counting up from 0, the first few binary numbers are as shown in Table 6.1, with the equivalent regular decimal numbers alongside.

You can see that any decimal number has an equivalent binary number, and it is easy to convert between the two systems. You will also notice that each time we add one more digit column to the binary number, it allows us to double the equivalent decimal number, and similarly, double the amount of information that can be coded in binary.

Bits and Bytes

Each binary number consists of a series of 0s and 1s—known as *binary digits*; these words have been combined into the single term *bits*. The abbreviation for "bit" is "b" (and the abbreviation is case-sensitive). Like decimal numbers, which have names for larger units as decimal places ascend (e.g., hundred, thousand,

TABLE 6.1
Binary and Decimal Numbers

Binary	Decimal
0	0
1	1
10	2
11	3
100	4
101	5
110	6
111	7
1000	8
1001	9
1010	10

million, etc.), so too do binary numbers have names for larger units. The next larger binary unit after the bit is known as the *byte*, which is a unit two binary orders higher than the bit (like the hundred is to the one in decimal numbers). Therefore, a byte equals 8 bits ($2 \times 2 \times 2 = 8$). The abbreviation for "byte" is "B" (upper case).

Units for Data Transmission and Storage

When binary data is sent from one place to another, an important consideration is how fast the bits are transmitted. This is measured in bits per second (bps), thousands of bits (kilobits) per second (kbps), millions of bits (megabits) per second (Mbps), or billions of bits (gigabits) per second (Gbps)—again, note the case-sensitivity of these terms and their abbreviations. (Sometimes these units are specified with a slash in lieu of the "p" for "per," as in b/s, kb/s, Mb/s, etc.)

While bits per second and its multiples are used for data transmission, when data is *stored*, the usual measure of capacity is quoted in *bytes* (B), again with the appropriate prefix added to denote unit multiples such as thousands, millions, etc. (e.g., kB, MB, GB)

Note, however, that binary data follows the powers of two (as befits its binary nature), not the powers of 10 as does the decimal system. Therefore, traditional practice assigns the nearest power of two to the terms we have developed for decimal counting. For example, the kilobyte (kB) unit is slightly more than one thousand bytes, because it actually refers to 1024 (2^{10}) bytes. Similarly, the megabyte (MB) is actually 1,048,576 (2^{20}) rather than an even million bytes. There is some dissent among practitioners on this point today, however, with some defining these terms using the exact decimal values (i.e., 1 kB = 1000 bytes) rather than the binary values for simplicity. Note also that the lower case "k" is traditionally used when abbreviating "kilo" prefixes, to avoid confusion with another scientific term (Boltzmann's constant) that is denoted by the upper case "K," whereas all other prefix abbreviations use upper case letters. (This practice is also not universally observed among today's practitioners, however.) See Table 6.2.

Data Bandwidth

Somewhat confusingly, the rate at which data can be sent over a particular distribution or transmission medium is often referred to as its *bandwidth*. Note that this is not the same as the analog bandwidth of an RF channel referred to in Chapter 4. There is no fixed relationship between the analog bandwidth of a channel and the bit rate that can be transmitted and received through that channel. The latter depends very much on the method of coding and modulation used for carrying the data (up to the limits imposed by fundamental laws of physics and mathematics).

TABLE 6.2
Data Storage Unit Multiples Explained

Unit Name	Order (Binary, Decimal)	Abbreviation	Actual Binary Value
Byte	Unit (2^0, 10^0)	B	1 byte
Kilobyte	Thousand (2^{10}, 10^3)	kB	1024 bytes
Megabyte	Million (2^{20}, 10^6)	MB	1,048,576 bytes
Gigabyte	Billion (2^{30}, 10^9)	GB	1,073,741,824 bytes
Terabyte	Trillion (2^{40}, 10^{12})	TB	1,099,511,627,776 bytes
Petabyte	Quadrillion (2^{50}, 10^{15})	PB	1,125,899,906,842,624 bytes
Exabyte	Quintillion (2^{60}, 10^{18})	EB	1,152,921,504,606,846,976 bytes
Zettabyte	Sextillion (2^{70}, 10^{21})	ZB	1,180,591,620,717,411,303,424 bytes
Yottabyte	Septillion (2^{80}, 10^{24})	YB	1,208,925,819,614,629,174,706, 176 bytes

COMPUTERS

Personal Computers

The personal computer, or PC, is familiar to most people. It is a general-purpose machine that can process large numbers of bits quickly to carry out many different tasks, depending upon the software loaded on it. The main hardware components include all or most of the following:

- Central processing unit (CPU)
- Random access memory (RAM)
- Mass storage: disk drives, optical drives, and so on.
- User interface: keyboard, mouse, picture monitor, loudspeakers
- Graphics subsystem
- Sound subsystem
- Modem for analog telephone line interface (if needed)
- Network interface(s) (if needed)
- Other physical hardware interface(s)

The various components may be combined into one compact unit, as in a laptop computer, or may be in several individual components, as in a desktop computer. In broadcast studios, computers may also be built into *rack-mount* units for convenient mounting in equipment racks.

Computer software includes the *operating system* needed to run the basic computer functions and provide the foundation that allows other software, called *applications*, to work. Applications are programs loaded onto the computer to perform functions such as word processing or more specialized tasks for broadcast operations. These days, most software is delivered from the supplier on CD-ROM discs or downloaded over the Internet.

Broadcast stations use PCs for many purposes directly associated with program production and sending to air, including processing, recording, and editing audio and video; producing computer graphics and captioning; and controlling and monitoring other systems and equipment.

Servers

Servers are powerful computers with a particular function on a *network*. They provide a centralized location for various network management and operational functions. For example, a server may run an application, such as a database, that can be accessed by many *clients* (i.e., other users on the network, typically accessing the server via a regular PC). File servers provide central storage of data for clients. Other servers manage access to the Internet or e-mail systems for network users. Video and audio servers are computers with particular capabilities dedicated to storing and streaming video and audio programming.

Clients

Clients are actually software elements running on distributed PCs that are connected to a network, which work in conjunction with a server. When a user launches a client on a PC, it runs *locally* on that PC, but it will interact with a server to send information to, or request information from, resources stored on that server or other remote locations. For example, when a user wants to read e-mail, he or she launches an e-mail client on his or her PC, and that client displays e-mail messages on the PC that are actually stored on a remote server.

Specialized Computers and Processors

Although PCs fill many roles in broadcasting, some applications require more power and/or very high reliability. In some situations, it may be possible to meet the requirement by using combinations of multiple PCs working together. In other cases, high-performance computers with specialized processors and/or operating systems are used.

Many items of digital broadcasting equipment incorporate *embedded* computer-type processors (built into the equipment itself), with instruction sets held in *nonvolatile* (permanent) memory, and which may not require an operating system or application program. The advantage of embedded systems is that they can often be made more reliable than general-purpose computers, and performance can be tailored to the particular application.

STORAGE

Files and Folders

When a computer needs to store data (e.g., programs, text, audio, video), it packages it in the form of an electronic *file*. Files come in different types, with a different *extension* (usually a three- or four-letter suffix following a period) on their name to identify them (e.g., ".doc"). To organize them and make them easy to find, files are usually placed in *directories* or *folders* in the storage device being used. They can also be transferred from one computer to a storage device on another computer via a private network or the Internet or copied to removable media such as an optical discs or solid-state memory devices, which are described later.

Disk Drives

All PCs, and many other computers, use *hard disk drives* (also known as a *hard disk* or *hard drive*) for long-term storage of programs and data. With hard disk storage, the data can be rapidly accessed and will not be lost when the computer is turned off. Such drives can store up to terabytes of data on one or more spinning magnetic disks or *platters* (most hard drives have multiple platters on a single spindle). Each disk has thousands of concentric tracks that can be rapidly accessed by record/replay *heads* that move in and out over the surface of the disk. They do not actually touch the surface of the disk, because there is an air cushion between, so the disks do not wear out due to friction with the head, as happens with magnetic tape storage.

Multiple hard disk drives may be combined together in different ways to achieve much higher reliability, and/or greater capacity, than can be achieved with a single disk. One such arrangement is known as RAID (*redundant array of independent disks*). These arrays are generally used in conjunction with servers. Large-scale disk storage systems may be in boxes separate from the associated server computers.

Other types of drives that have removable magnetic disks, such as the floppy disk, are sometimes used for data storage. Their capacity is much less than hard drives, and they have largely been replaced by optical disks and solid-state storage.

Tape

Where even larger amounts of data have to be stored for archive purposes, the computer (usually a server) may transfer it to special magnetic data tapes. These can store very large amounts of data for long periods with great reliability. The disadvantage is slow access to any particular piece of data, because the data is stored on a long ribbon of tape that may take a considerable time to shuttle to the required location.

Optical Discs

An alternative to magnetic disk or tape data storage is *optical disc* recording, based on the *compact disc* (CD), *digital versatile disc* (DVD), and *Blu-ray disc* (BD) recording formats (note that optical discs are generally spelled with the letter "c," whereas magnetic disks use a "k"). Introduced initially as read-only memory (ROM) devices, these are available in various versions that can be recorded only once ("write-once" formats) or that can be both recorded and erased, for multiple uses ("erasable" or "re-writable" formats).

CDs, DVDs, and Blu-ray discs look similar, but CDs can hold about 700 MB of data, while DVDs can hold 4.7 GB on *single-layer* formats and 8.5 GB in *dual-layer* formats. As the name implies, dual-layer formats store a second layer of information above a previously recorded layer, but the upper layer remains transparent so the lower layer can still be read. (Note the use of the plural in referring to recordable DVD formats, because there are two different varieties of write-once DVDs with certain incompatibilities—DVD-R and DVD+R—although most DVD drives used in current computers can handle either format interchangeably.) The erasable variety of DVD is called DVD-RW, which can store up to 4.7 GB. Some write-once DVDs are made in *dual-sided* formats, which double the capacities cited.

Blu-ray discs also come in either write-once (BD-R) or erasable (BD-RE) varieties, and with one to four layers of recording (BD-RE is only capable of three layers at this writing). Capacities are 25 GB for single-layer discs, 50 GB for dual layer, 100 GB for triple layer, and 128 GB for quadruple-layer BD-R.

Solid-State

The disadvantage of all the previously mentioned storage systems is that they involve moving mechanical parts that can wear out or break. In addition, the speed of storage or access to individual items of data is limited by the physical speed of access to the media. Both of these drawbacks are overcome with *solid-state* storage. This is memory based on silicon chips, comprising integrated circuits with millions of transistors and other components inside. Modern computers have always used solid-state storage for their RAM functions, because it allows the currently running processes to run quickly. Traditional solid-state memory has several drawbacks, however: It loses data when power is turned off, has limited capacity, and, for many years, was relatively expensive compared to alternative storage technologies.

More recently, solid-state memory chips have been produced with much greater capacities, which have reduced their cost-per-byte, although access times have slowed somewhat on these large chips. While less appropriate for RAM applications due to their speed, this type of solid-state memory is suitable for long-term mass storage because it can also be made *nonvolatile* (i.e., it does not lose data when the power is removed) and can be produced in a removable card or other small device form. Examples are the *flash memory* cards used for digital cameras and USB "stick" devices. This type of memory is also used for storage on mobile and handheld devices like smartphones and tablets and is increasingly available (often as an option) in laptop computers, where despite the speed issues noted above, it still delivers much higher-speed access to data than the hard disks traditionally used for storage in laptops. Solid-state storage also has the advantage of quieter operation and insensitivity to motion, which can sometimes be advantageous when laptops are used in broadcast applications. The higher long-term reliability of solid-state storage is also a plus for mission-critical operations.

COMPUTER NETWORKS

LANs and WANs

Computer networks are used to connect two or more computers together to allow them to share data or resources. Frequently, one or more servers may be connected to multiple client computers on a network, and devices such as printers may also be connected for shared access. There are two main types of network: *local area networks* (LANs) and *wide area networks* (WANs). As might be expected, LANs generally operate over comparatively small areas, such as within a building, whereas WANs operate over much larger distances, spanning up to whole countries or the world.

Ethernet

By far the most common form of LAN is known as *Ethernet*. It operates over various forms of cabling: twisted pair (similar to telephone wires), coaxial cable, or fiber-optic cable, and at various data rates, ranging from 10 Mbps to over 1 Gbps. Other network technologies are occasionally used for specialized purposes with LANs and WANs.

WiFi

While Ethernet is almost universally used as a wired LAN computer network, an equally popular wireless LAN system today is *WiFi*, officially known by its standard name of IEEE 802.11. Introduced in the late 1990s, there have been many variations of WiFi to emerge since (each adding a different suffix to the 802.11 nomenclature, such as *802.11b* or *802.11g*), which have increased the maximum interconnection speed possible for a user on the network from the original 1 Mbps to, at this writing, maximum speeds approaching 1 Gbps.

Like Ethernet, WiFi access is typically built into many computing devices today as standard equipment, using an antenna (usually embedded into the device, i.e., invisible to the user) in lieu of a wired jack that is used for Ethernet connectivity. When a user of such a device wishes to connect to a computer network wirelessly, the WiFi interface on the device seeks out available WiFi *access points* in the vicinity. These access points are wireless transceivers that can wirelessly connect to multiple WiFi users and route their network connections to a broadband network to which the access point is also connected, typically through a wired interface. Thus, the WiFi access point acts as a networking *hub* that connects many individual wireless users to a wired computer network.

WiFi is a key component of today's burgeoning handheld and mobile computing space and is used in both private and commercial applications. Based on its design as a wireless LAN, it has distance limitations of approximately 100 to 200 feet indoors or about two to three times that outdoors. This means that a WiFi user must be within those distances from an access point to be able to connect to the network wirelessly.

Broadcasters frequently use WiFi in their daily production operations, and increasingly, audiences access broadcasters' content using WiFi to connect to streaming media via the Internet, as explained below.

Internet

The Internet is the worldwide, interconnected system of networks that allows connected computers to potentially communicate with any other connected computer or server. It consists of enormous data *backbones* transmitting billions of bits of data per second, with smaller distribution systems attached as tributaries to the backbones.

An *intranet* is a private network, which uses similar software and protocols to those used with the Internet, but which is not accessible to the general public. Intranets may be established for local access only (for example, within a company's corporate premises), or they may be available in multiple locations via secure paths over WANs.

The Internet is increasingly used by broadcasters today as an alternative to over-the-air delivery of content to audiences. For more on this topic, see Chapters 14 (for Radio) and 17 (for Television).

Protocols

When computers communicate over a network, or between networks, they must use an agreed-upon *protocol* (which establishes the data format and operating rules) for data communication, if they are to understand each other. Numerous protocols have been established for various purposes, but the most widely used today in computer networking is called *IP* (*Internet protocol*). IP is used not only for communication over the Internet but also for many other networking purposes, sometimes in conjunction with other protocols.

One of the key features of IP is that every device using it is assigned a unique *IP address*, which allows messages to be routed to and from the device. Each data stream in an IP network is broken into specific pieces of the data, called *packets*. Each packet has its own unique identifier containing information about where it came from and where it is intended to go.

Another protocol commonly used in conjunction with IP is *TCP* (*transmission control protocol*), giving rise to the combined term *TCP/IP*. An important feature of TCP/IP is its ability to adapt to a wide variety of network conditions. On a network with reduced bandwidth or under congested conditions, TCP throttles back the IP packet rate to a lower—but still robust—transmission speed, and when conditions allow, it accelerates to the maximum rate possible for improved performance.

Switches and Routers

LANs are made up of interconnected client and server computers, frequently connected together in a "hub-and-spoke" arrangement, where each computer is connected back to a central *network switch*. This enables the computers to communicate with each other using a suitable protocol.

Routers can also serve as a switch but, in addition, they provide capabilities for secure control over what messages on the LAN are routed to which IP addresses. They allow rules to be established for which computers or servers are accessible to each other on the LAN and for connections to other networks and the Internet.

Security

A *firewall* is usually provided for a LAN to prevent unauthorized access to clients or servers from computers outside the network. The firewall may be a hardware device associated with the gateway from the outside world or may be software based, running on a particular server or client PC.

Critical messages and data sent to systems outside of the firewall are frequently sent with *encryption*. This means they can only be decoded and understood by users with authorized computers.

Internet Streaming

Streaming is the blanket term for the distribution of audio or video, plus supplementary data, from a central distribution point over the Internet to multiple users. It is important to recall that computer networks, and the Internet in its earliest form, were never intended to provide real-time audio or video content. They were intended as systems that allowed computers to transfer files to one another and could do so at whatever pace the traffic on the network allowed (like sending a file from a PC to a networked printer, for example). Using such a computer network for service more analogous to broadcast delivery of real-time audio or video media was a big step and required some highly specialized retrofitting.

Technology

Traditional broadcasting is a point-to-multipoint distribution system: one transmission site serves many thousands or millions of people in a given geographic area, with no limitations so long as they are within the range of the over-the-air signal. Internet streaming is different: it takes one digital audio or video program and copies it many times over to reach multiple customers individually via computer networking. Each additional user requesting the program requires additional equipment and networking resources to be provided for distribution.

In very simple terms, the program produced by a station, or some other source, is fed into a *streaming media encoder*, which digitally compresses the video and / or audio and encodes it into a form that can be efficiently distributed over the Internet. There are numerous streaming media encoding systems, called *codecs* (for coder-decoder), which take a broadcast signal and reduce its bit rate for efficient delivery over the Internet. These codecs are critical elements of the streaming process, because they reduce bit rate by a large margin—a factor of 10:1 or more is not uncommon—and unlike traditional data compression algorithms, they do so in a *lossy* (as opposed to a *lossless*) way, such that the original file is <u>not</u> reconstructed bit-for-bit upon decoding or playback. Nevertheless, this digital compression (or more properly, "bit-rate reduction") is performed in a way that is relatively imperceptible to the user because these codecs use technologies that exploit various insensitivities of human perception. Without such techniques, the quality of the program would be significantly reduced when compressed to such a reduced bit rate, but human listeners and viewers generally do not notice the substantial amount of data that these codecs eliminate. They are therefore referred to generally as *perceptual codecs*, and they contribute greatly to the efficiency and thereby to the viability of streaming media service on the Internet.

Streaming media encoding can be done in two ways: (1) in hardware, where the encoding is done by a physical chip or (2) in software, where the software does the encoding on a general-purpose platform. Hardware encoders tend to be faster but typically require an upgrade or exchange of a physical chip to update encoding capabilities, whereas software decoders are generally slower but offer greater ability to easily update tool sets. See Chapter 15 for more information on data compression.

Once a stream has been created, it must be fed to a server, which then copies the stream for each of the customers who are asking for the program. The service provider allocates a certain amount of bandwidth for the feed and provides the computer servers and data distribution equipment to get those feeds to the Internet for distribution to the end customer. If more requests for the feed are received than data space or program feeds have been created, then no more customers can be served.

Finally, the data streams must be transmitted over the Internet. Each data stream is broken into packets and distributed using an Internet protocol as described earlier. However, unlike regular broadcasting, there is no guarantee that packets will arrive in a particular order, or how long packets will take to reach their destination. Thus, if the Internet backbone or the Internet access connection to a particular user is under heavy load, the distribution of the packets for the audio/video stream may be disrupted. This can produce jumps in the picture or skips in the audio when a user is listening to the streaming content, or simply termination of the stream to that user. As broadband Internet service improves and is increasingly deployed, such difficulties will decrease, but some finite capacity limits will theoretically always apply to any computer network connection, especially when wireless access is involved. Such limits do not apply to broadcast service. The restrictions mentioned previously on how many customers can be served particularly apply to Internet distribution using the *unicast* model, which is by far the most common. A technique called *multicast* is also possible in IP networking, which can reduce such problems. Multicast makes a single stream available to many users by replicating it within the network. IP multicast requires many changes in the way the Internet interconnections and routing are set up, however, and not all service providers offer multicast capability. So in many cases, not every segment of the network between a given content source and its users will be configurable in multicast mode, so unicast distribution will still be required for all or part of the network paths involved.

Data Rate and Signal Quality

The data rate of the stream is governed by whether the content is audio, video, or both, and by the perceptual codec used. It may also depend on the intended final delivery link, for example, whether the user is requesting the stream for a high-speed broadband connection to the home or for a wireless connection to a smartphone. The ultimate quality associated with a given bit rate depends mostly on the type of encoding used—the better the encoder, the better the reproduced signal quality for a given data rate. In any case, audio-only streams require much lower bit rates to deliver high-quality sound than video or video/audio streams require for high-quality images.

For audio, the most common data rates are 8 to 24 kbps for voice-quality sound (e.g., communication/telephony applications) and 48 to 128 kbps for high-fidelity sound (e.g., music downloads or streams). Note, however that these values or current at the time of this writing, and codecs continue to improve. It is likely that these values may see ongoing reductions over time.

Video Standards

Unlike conventional television, the devices that are used to view streaming video are not single-standard devices with respect to video formats or even screen sizes. It is therefore difficult to know exactly what a given video stream will look like on every target device, or even if it will play at all, because not all devices are equipped to decode every video encoding format. For this reason, many services provide video stream of the same content in multiple formats and quality levels, allowing the user (or the user's device) to select the most appropriate offering for the user's current viewing conditions.

CHAPTER 7

Radio Frequency Waves

This chapter describes some of the scientific principles of radio waves that are fundamental to all types of broadcast transmission.

ELECTROMAGNETIC WAVES

It is fairly easy to understand that sound waves can be carried through the air, just as waves move on the sea. Sound waves are therefore a mechanical type of wave called *longitudinal waves*, which exist solely as disturbances within a medium. The medium can be of any state—solid, liquid or gas—but if no medium is present, this type of wave cannot exist (which is why there is no sound in a vacuum, for example). Light, on the other hand, is a so-called *transverse wave*, which exists independent of the medium through which it passes. Hence, light can pass through the vacuum of space, as well as through air, water, and so on. Light can also be considered as part of a larger class of transverse waves called *electromagnetic waves*. In simple terms, electromagnetic waves are vibrations of electrical and magnetic energy that can travel long distances, even through the vacuum of space or, to a variable extent, through other materials. They are the foundation of over-the-air broadcasting.

Types of Electromagnetic Waves

Electromagnetic waves can exist over a wide range of frequencies, And depending on their frequency, they may have very different properties. The following broad classifications make up the electromagnetic radiation *spectrum*. Starting from the lowest frequencies, they are as follows:

- Radio waves
- Microwaves (very short radio waves)
- Infrared waves

- (Visible) Light waves
- Ultraviolet waves
- X-rays
- Gamma rays

Radio waves are usually referred to as RF, standing for *radio frequency*. In the early days of radio, the term *ether* was used to describe the undetectable medium that was then thought to carry electromagnetic waves. Later experiments proved that the ether did not actually exist and no medium was required to carry the waves (which, as stated above, are a form of energy unto themselves), but occasional references are still made to radio as "traveling through the ether."

Frequency, Wavelength, and Amplitude

As stated earlier, radio waves come in a range of frequencies. Another way of saying this is that they exist over a variety of *wavelengths*. Frequency and wavelength are really statements of the same thing, i.e., the rate of oscillation or "vibration" of a radio wave, but the two terms specify it in a manner that is the exact inverse of each other. As frequency goes up, wavelength gets smaller, and vice versa. Wavelength, as the name implies, is quoted in terms of linear distance, generally in metric units (e.g., meters). Frequency, on the other hand, considers the time it takes for a radio wave to perform its oscillation, and specifies it as the number of oscillation periods a given radio wave completes in one second. It was originally measured in the straightforwardly titled standard unit of "cycles per second" (cps), but this value is now quoted with the unit *Hertz* (Hz), in honor of Heinrich Hertz, the German physicist generally considered the first to conclusively prove the existence of electromagnetic waves. As with other physical units, metric multiplier prefixes are appended to the Hz abbreviation for higher values, such kHz ("kilohertz") for thousands of cycles per second, MHz ("megahertz") for millions of cycles per second, and GHz ("gigahertz") for billions of cycles per second.

Occasionally a certain radio channel or contiguous group of channels will still be referred to today by wavelength (e.g., "the 10-meter band"), but it is nearly universal current practice to describe a particular part of the electromagnetic spectrum by its frequency value instead (e.g., "28,000 to 29,700 kHz"). Frequencies used for broadcasting in the United States range from about 500 kHz to 12 GHz.

Whatever their frequency, all electromagnetic waves travel through a vacuum at the speed of light (about 186,000 miles per second, or 300 million meters per second). This speed is almost the same in air, but it decreases in

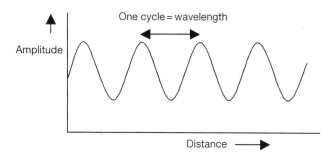

FIG. 7.1. Radio Wave

other materials. Figure 7.1 represents a radio wave as it travels through space; as described above, the *wavelength* is the distance that one "wave" or *cycle* of oscillation occupies in space. It is easy to calculate the wavelength of a signal if you know its frequency, or vice versa, because the two values multiplied together always equal a constant number—the speed of light. Therefore, the wavelength of a given radio wave in meters is approximately 300 million divided by the radio wave's frequency in Hertz.

As shown in the figure, radio waves also exhibit a certain *amplitude*, which is a depiction of their intensity (or *level*). It is illustrated by the vertical dimension in the figure. Thus these two independent variables—frequency and amplitude—are identifying characteristics of any radio wave.

FREQUENCIES, BANDS, AND CHANNELS

Radio frequency transmissions are divided up into contiguous *bands* for different purposes. These groupings are usually due to the different physical characteristics or behavior of radio waves at different frequencies (more on this below). To understand the properties of different radio waves, let's first calculate the wavelength of some typical radio frequencies found in each of the main broadcast bands. As explained above, this is performed by dividing the speed of light by the frequency of interest:

Band: MF – Medium Frequency – AM (*amplitude modulation*) Radio Band (535–1705 kHz)
Channel: AM Radio – 1120 kHz
300 million meters/sec ÷ 1120 thousand cycles/sec = 268 meters (880 feet)

Band: VHF – Very High Frequency – FM (*frequency modulation*) Radio Band (88–108 MHz)

Channel: FM Radio – 98.1 MHz
300 million meters/sec ÷ 98.1 million cycles/sec = 3 meters (10 feet)

Band: VHF – Very High Frequency – Television Band (54–216 MHz)
Channel: VHF TV, Channel 8 – 183 MHz
300 million meters/sec ÷ 183 million cycles/sec = 1.64 meters (5 feet)

Band: UHF – Ultra High Frequency – Television Band (470–806 MHz)
Channel: UHF TV, Channel 40 – 629 MHz
300 million meters/sec ÷ 629 million cycles/sec = 0.48 meters (19 inches)

Band: SHF – Super High Frequency – Broadcasting Satellite Ku Band
(11–14 GHz)
Channel: Direct Broadcast Satellite, Transponder 30 – 12.647 GHz
300 million meters/sec ÷ 12.647 billion cycles/sec = 2.37 centimeters (1 inch)

As this exercise illustrates, the wavelengths of different broadcast bands vary quite a bit, from hundreds of feet to an inch or less. Because of these differing wavelengths, there are major physical differences in the design of antenna types used to transmit and receive signals in different bands. The amplitudes (or *power*) of these signals also greatly influence the design and size of these antennas.

Each radio band is divided into individual channels, and each of these channels includes a range of frequencies. The range of frequencies included in a channel from lowest to highest is known as the channel's *bandwidth*. (The term may also refer to any particular range of frequencies, not just those in RF.) For simplicity, however, a channel is often identified by its *center frequency*, so that only one numerical value (rather than two) will have to be cited when referring to a particular channel. If the center frequency is given and the channel bandwidth is known, the upper and lower frequency limits of the channel can be easily derived. For example, a channel with a 2 kHz (2,000 Hz) bandwidth centered at 100 kHz occupies the spectrum between 99 kHz and 101 kHz.

The UHF and SHF bands have further subdivisions, with bands that are used for terrestrial radio links, satellite links, and for satellite broadcasting. These include the L, S, C, X, Ku, K, and Ka bands, with frequencies ranging from about 1 GHz to 40 GHz.

Propagation Properties

Because of the varying wavelengths, waves in different bands have different propagation properties. In particular, the shorter its wavelength, the more a wave tends to travel in straight lines and to be blocked by obstacles in its path. Longer waves, such as those found in the AM medium frequency band, tend to "flow" (or *diffract*) around obstructions, and may propagate in multiple ways, such as via a *ground wave* and/or a *sky wave*. See Chapter 20 for details.

Like any wave, broadcast waves also experience *attenuation* over distance (or *path loss*), by which a transmitted signal gets weaker and weaker as it travels into space away from its transmission source. Thus, during such wave propagation, as the broadcast wave travels through space, its frequency stays the same (for practical purposes), while its amplitude decreases. At some point, every transmitted broadcast signal becomes too weak to receive—just as a sound becomes too quiet for us to hear after it travels a sufficiently long distance from its source.

Notwithstanding the statement above that a transmitted frequency remains constant, there are some special applications that take advantage of small changes in frequency of a transmitted wave that do occur when two objects are moving with respect to each other. This is known as the *Doppler effect*, or *Doppler shift*. Indeed, some applications, such as speed-detecting radars and radio astronomy spectroscopy, are based on this phenomenon. For most broadcasting applications, however, such frequency shifting can generally be ignored.

RF OVER WIRES AND CABLES

As described previously, RF waves travel through space as electromagnetic waves, but they also can be carried as varying electrical voltages over copper wires and cables, just like baseband audio and video signals. The distance they can travel over wires is limited by signal losses in the cable at RF frequencies, however, and this varies with the frequencies involved. Generally, the higher the frequency, the greater the loss per unit of distance, meaning that higher frequencies can travel for shorter distances over a given cable than can lower frequencies.

MODULATION

Carriers and Subcarriers

Broadcast signals are intended to carry sound, pictures, and other information over long distances. Therefore, the radio waves used for broadcast transmissions are chosen specifically for their ability to travel over desired distances before they are attenuated to levels that are no longer possible or practical to receive. These frequencies are different from the frequencies originally occupied by the sound, light, or data signals we want to transmit, so those *baseband* signals must be translated to different frequencies for broadcast. This translation process is called *modulation*, and the frequencies the baseband signals are translated to for broadcasting are called *carriers* or *carrier waves*. Carrier waves are the actual signals broadcast through the air, and as the name implies, they carry the baseband signals representing broadcast content within them.

The process of modulation can change the waves of the baseband signals in many different ways. In traditional analog broadcasting, modulation simply changes the baseband signals' amplitude and frequency to different carrier amplitudes and frequencies in a fairly direct translation process, but there are several other more complex and indirect modulation methods used for different broadcast purposes today, as discussed in the following sections.

First, a few other terms must be defined, however: A broadcast receiver picks up the transmitted radio waves via its antenna, and selects the desired RF channel with a *tuner*. The tuner sends the channel's carrier wave to a *demodulator*, which turns the signals back into baseband audio, video and/or data signals for presentation to the audience. This demodulation process is essentially the inverse of the modulation process described above. More detail on this can be found in Chapter 20.

You will also see the term *subcarrier* used below. It has two meanings. It may refer to a modulated carrier that is combined with other signals and then modulated again onto a higher carrier frequency. Subcarrier terminology may also refer to the use of multiple carriers of different frequencies that are used together in a single, multi-frequency transmission system (for example, COFDM, as mentioned later in this chapter).

Amplitude Modulation

In *amplitude modulation* transmissions, known as AM, the program signal is used to modulate (vary) the amplitude of the carrier wave, as illustrated in Figure 7.2. When the amplitude of the program signal is zero, the carrier remains unmodulated. As the instantaneous amplitude of the program signal increases up to its maximum, the carrier amplitude varies accordingly, up to the maximum amount—100 percent modulation.

Frequency Modulation

In frequency modulation transmissions, known as FM, the program signal is used to modulate the frequency of the carrier wave, as illustrated in Figure 7.3. The amplitude of the carrier wave remains constant throughout. When the

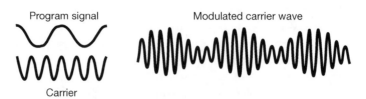

FIG. 7.2. Amplitude Modulation

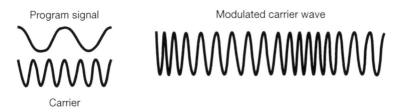

FIG. 7.3. Frequency Modulation

amplitude of the program signal is zero, the carrier remains at its nominal frequency. As the instantaneous amplitude of the program signal increases up to its maximum, the carrier frequency varies accordingly, up to the maximum amount allowed, which is usually 100 percent modulation.

Quadrature Amplitude Modulation

A variant of AM is called *quadrature amplitude modulation* (QAM). This provides a way to carry additional information in some types of radio and television transmission without using significant extra bandwidth. In particular, it is used to carry supplementary services in AM radio and, with a subcarrier, to carry the chrominance information in NTSC television.

With QAM, two separate signals are modulated onto two separate carriers that are of the same frequency, but one-quarter wavelength out of *phase* with one another, as shown in Figure 7.4. Phase is really a statement of time, but instead of being quoted in typical time units like seconds or fractions thereof, it is quoted in terms of wavelength of a particular frequency, usually in relative terms or by degrees of difference to some reference wave (e.g., "a quarter-wave out-of-phase," or "90 degrees out-of-phase"). Quadrature always refers to

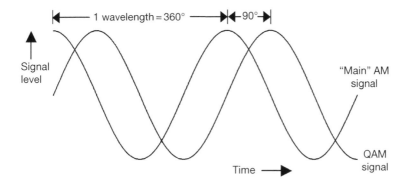

FIG. 7.4. Example of Two Waves in Quadrature

a relationship between two waveforms of the same frequency that are separated in time by one-quarter of a wavelength, or 90 degrees, as depicted in Figure 7.4.

The figure shows that when one carrier is at a positive or negative maximum level, the other one is always at zero (half-way between maximum and minimum), and there is always a fixed 90-degree phase offset between the two signals. This fact allows the two carriers to both be modulated with different information, and be separately demodulated at the receiver. Having the same frequency, they occupy the same portion of the radio spectrum. This basic function is utilized in more complex ways by higher order forms of phase-varying modulation techniques, as discussed below.

Digital Modulation Systems

Digital broadcasting still uses a carrier wave but with more sophisticated modulation methods used to vary the amplitude, frequency, or phase of the RF signal.

An important element to understand about digital modulation (and all the digital broadcast systems that are thereby enabled) is that the RF signal emitted into the air is actually an analog waveform. The continuously variable carrier wave generated by a digital modulation system is formed by digital techniques, however, and thus its waveform generally appears "step-like," in that it carries all its information at discrete modulation levels. Think of this as climbing a ladder versus pulling yourself up a rope. Using a ladder, you can only climb in specified intervals, whereas on the rope you can start and stop at any point along its length.

There are many ways in which this "step function" can be applied to an RF carrier wave via digital modulation. A few are described below, with more detail available in later chapters.

One such method is *Vestigial Sideband* (VSB) modulation, a digital amplitude modulation system. An eight-level form of it, called 8-VSB, is used in ATSC digital television. This method is discussed further in Chapter 16.

Several versions of QAM are used in digital modulation systems (for example, in AM IBOC radio and digital cable television). In this case, the two carriers in quadrature are modulated to various discrete levels that, taken together, represent a particular digital *symbol*. There may be different numbers of symbols (typically 16, 64, or 256), depending on the application.

Phase modulation systems applied in digital modulation include the various versions of *phase shift keying* (PSK) such as *QPSK* and *8-PSK*.

Pulse code modulation (PCM) refers to a quite common method of encoding sampled analog signals in digital form. It is often used in baseband digital representations of audio and video (for example, in compact discs), although PCM signals may also be modulated onto an RF carrier.

COFDM

Various systems for digital transmission use modulation of multiple carriers within a transmission channel to make the signal more robust, particularly so it can resist *multipath* reception conditions. Multipath occurs when a receiver encounters multiple reflections of the same transmitted signal (each arriving at the receiver at slightly different times) caused by nearby structures or terrain. It is particularly problematic for mobile reception because the reflections are constantly changing as the receiver moves through space.

One such system, *coded orthogonal frequency division multiplexing* (COFDM) is used by many of today's digital broadcast systems, including IBOC, DAB, DRM, DVB-T and –T2, ISDB-T, and DTMB. It is also used for *electronic newsgathering* (ENG) digital radio links.

A COFDM transmission is composed of thousands of separate subcarrier frequencies, each of which carries a relatively low-speed data stream, modulated by a digital modulation technique as described above. For example, one form of DVB-T uses 1,705 COFDM carriers within an 8 MHz-wide channel, with each of the carriers modulated by QPSK. The same program information is typically modulated onto several different carriers, and often transmitted repeatedly at slightly different times, so that if the receiver momentarily fails to receive some of the carriers (for example, due to multipath interference), the program signal is not lost because the same data is redundantly carried elsewhere in the channel and/or repeated at a slightly different time. Such techniques are forms of what is generically referred to as *channel coding*, or more specifically, *interleaving* and *forward error correction* (FEC), which add robustness to all digital transmission and storage systems.

Sidebands

When an RF carrier (or subcarrier) wave of a particular frequency is modulated with a signal of another frequency, the resulting RF signal has additional RF components called *sidebands*. Sideband frequencies are typically the carrier frequency, plus or minus a whole multiple (1×, 2×, etc.) of the modulating frequency. It is these sidebands produced by modulation that carry the actual information in the radio wave.

As an example for AM, a carrier wave of, say, 650 kHz, amplitude modulated with an audio signal of 10 kHz, will produce a composite signal comprising the original carrier of 650 kHz plus an upper sideband of 660 kHz and a lower sideband of 640 kHz. The number and size of the sidebands depends on the type of modulation, but they always extend both above and below the original carrier frequency. In some cases, it is possible to transmit only part of the sidebands and still recover the program information at the receiver. In other cases, the main carrier is suppressed and only the sidebands are transmitted.

This concept of sidebands is important because it affects the amount of radio spectrum, or bandwidth, used for a given type of transmission. This is discussed later in the chapters on broadcast standards.

Light Modulation

For completeness, it should be mentioned that light waves can also be modulated to carry information. The most common applications are for optical sound tracks on film, and for carrying audio, video, or data over fiber-optic and laser links.

STUDIOS, PRODUCTION, AND PLAYOUT FACILITIES

CHAPTER 8

Radio Studios

This chapter explains the permanent facilities used for making radio programs, from the microphone in the studio to the equipment that feeds the program to the transmitter. Arrangements and equipment for the temporary facilities used in "remote" broadcasting from outside the studio are covered in Chapters 10 and 11.

TYPES OF STUDIOS

Radio studios are special rooms where audio programs or contributions are produced and prepared for broadcasting. Apart from their technical equipment, the important thing about radio studios is that they are designed or adapted to be isolated from outside sounds, so there is no background noise interference with a program. Further, they have special *acoustic treatment* on the interior surfaces of walls and ceilings to control the amount of *reverberation* (the persistence of sound due to multiple reflections after the sound source has stopped), so it does not sound as though a speaker is in a very *live* environment like a bathroom.

Local radio stations have one or more studios; larger numbers of studios are typically located in radio network centers, production centers, and elsewhere for remote contributions. There are several varieties of studio, depending on the type of content involved, and whether they are primarily for live, on-air use or for recording. A major distinction is whether the studio has a separate control room or whether there is a combined, or *combo*, studio and control room. In a combo facility, the *control board* operator is also the presenter or disk jockey (DJ), who does both tasks simultaneously. The combined arrangement is common for on-air studios today.

A studio with a separate control room is used primarily for more complex productions. The presenters and guests, or artists (known as *talent*) are in the studio, and one or more technical operators, often with other production staff, are in the acoustically isolated control room with the technical equipment. Usually there is a soundproof window between the two areas for visual communications. In some cases, a third room (acoustically isolated, but visually coupled via soundproof windows) may be included in the studio suite, for additional functions such as call screening for talk-radio shows, or for a separate announcer (often called an *announce booth*). Larger radio stations and network centers may also have a studio intended for live music or other performances, which typically is designed with more space in both the studio and control room, and generally houses more equipment.

At the other extreme, many functions for audio production today that previously needed a room full of equipment can be done on a personal computer, with only a few items of external equipment required. This development has allowed new methods of content creation to flourish, and has blurred the definition of what constitutes a radio studio. For this chapter, we will concentrate on the on-air type of studio used by most local radio stations, but the basic concepts discussed will apply to other studios as well.

A small radio station may have just one studio—the on-air studio—but most will have several studios for different purposes, perhaps including a news studio and a production studio for recording programs or commercials.

STUDIO OPERATIONS

Traditionally, a radio station produces on-air programs with one or more live presenters and guests in a studio at the station, playing music and other program segments from various devices in the room, and putting phone calls or other outside contributions on the air. When needed, the on-air studio operator may select audio from another source in the building (e.g., a newsroom), or from a remote source, such as a network feed or a sports venue, often mixing that audio with local sources in the room into the on-air program. In some cases, whole programs may come from an external network or program-syndication source, via a live feed or from a recorded medium. In this case, the station acts as a *pass-through* for such external content, as opposed to when it generates its own *locally originated* content.

For either model, the world of radio broadcasting is now largely automated, and in some cases, whole programs or *dayparts* may be broadcast without the intervention of a live presenter or operator at the local radio station.

Automation and Live Assist

The introduction during the late 1980s and the 1990s of computer hard drive audio storage equipment and computer-controlled automation for radio (described later in this chapter) changed the way most stations operate. Previously, it was largely a manual operation to play music, other material or program segments, and advertisements from different types of tape decks, records or CDs, interspersed with live introductions and announcements. Now most material is stored as *audio files* on a PC, dedicated *hard disk recorder,* or server, and these files are played back under the control of *program automation* software. The audio files in this assortment of content may include recorded linking announcements made by a local (or remote) presenter, called *voice track-ing,* and for some types of programs, the operation can be run essentially unattended for long periods of time.

Even when the program format or station preferences require the presenter to talk live on-air (e.g., for phone-in shows), the computer can still handle the playout of all the other program segments. This mode is called *live assist,* in which the computer automation system pauses when it is programmed to expect a live break from the announcer. When the announcer completes the live break, he or she tells the computer to resume, and the automation system continues to present pre-programmed content until it encounters the next live break.

Remote Voice-Tracking

Remote voice-tracking takes the process one step further. In this case, a presenter or DJ records all of the voice segments for a radio program in advance, either in the station or at a remote location (perhaps in his or her home), and stores the recordings as audio files on a workstation or computer. If done remotely, the files are then transferred (over the Internet, a private network, or a phone line) to the hard drive storage at the local station. The segments are identified in the automation *playlist* and are played out on-air seamlessly with the music, just as though the DJ was live in the studio during the broadcast.

Similar techniques allow station groups to share programming among different stations, and in some cases, to control and monitor operations from centralized remote locations. The latter is known as *centralcasting.*

Ingest

An important operation for most radio stations is *ingest.* This involves receiving program material from outside sources and preparing it for use by the station. Feeds from networks frequently come in on *satellite links,* and live feeds may be fed to the on-air mixing console, as shown in Figure 8.1. Other material may be fed to a dedicated recording device or to a storage location on the station's computer network for integrating into later programs. Material may also be moved as audio files between workstations or computers within the radio station, digitally copied from CDs (*CD "ripping"*), or downloaded from external sources via the Internet or other network sources, and thus the ingest workstation may be a busy system that is best located outside of or separate from the on-air studio.

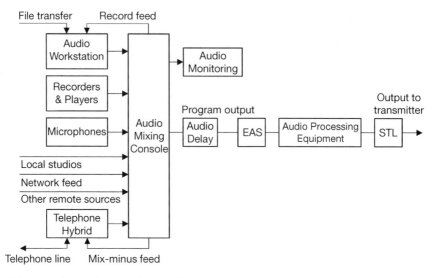

FIG. 8.1. Radio Studio Air Chain Block Diagram

Complete program segments also may be produced for the station at a different geographic location and transferred to the station over a private data network or via the Internet.

Editing

At one time, most audio editing performed at radio facilities to prepare news stories or other material for broadcast was done on audio tape machines by physically cutting and rejoining the tape, using razor blades and adhesive tape. In the 1990s, however, radio producers began transitioning to doing their audio editing on hard disk–based *digital audio workstations* (DAWs), some of which were built upon PC or Macintosh computers, while others used proprietary hardware. Today this function is routinely performed on any typical PC, using inexpensive software designed for the purpose.

Audio Storage

Nowadays, nearly all radio stations have their recorded music, commercials, and other material stored on computer hard drives. The continually increasing convenience of this type of storage and its continually decreasing cost per megabyte have made this possible. Advantages of hard disk storage and playout systems, compared to previous analog systems, include the following:

- Increase in audio quality
- Much reduced wear and tear, plus increased reliability

- Easy automation with a single computer program on a single machine
- Automatic creation of the program log
- Integrated storage and playout of content metadata (e.g., text of song title and artist name)
- More efficient use of station personnel

SYSTEM CONSIDERATIONS

Stereo or Mono

Most current audio equipment and studios are designed to work in two-channel stereo mode, and virtually all FM stations produce stereo programming. AM stations that transmit in mono only may also be set up to produce stereo programs, but they may combine the left and right channels to a mono signal at the output of the studio chain. Older stations may use only mono equipment (or stereo equipment in operated in mono mode) with equipment interconnect wiring done using single mono channels. When AM stations with mono studios add IBOC transmission, they often upgrade their studio facilities and operations to stereo because AM IBOC is a stereo service.

Analog, Digital, or AoIP

The trend in radio broadcasting is to replace equipment based on analog signal processing with digital equipment. There are many advantages to this change, including those noted above for hard disk–based storage and a general increase in capabilities. Some stations may still use analog audio mixing consoles and analog interconnections, however, either because they have serviceable legacy installations or because they choose not to upgrade their whole facility to digital. Most items of digital audio equipment can be used as digital "islands" in an overall analog system, using analog-to-digital and digital-to-analog converters. As mentioned in Chapter 5, every analog-to-digital and digital-to-analog conversion introduces a small quality loss, so the number of conversions should be kept to a minimum. New facilities are almost always built as fully digital systems today, and it is becoming increasingly difficult to even find analog audio hardware (other than microphones and speakers) in today's professional marketplace.

Note that even within digital systems at radio stations there are two classes of systems today. One is the earlier "discrete" or "parallel" digital approach, in which individual devices maintain individual input and output connections to and from one another, like they do in analog systems. The main difference between them is that instead of analog audio wiring between devices, a digital interface—typically AES3—is used for interconnection. The other, newer approach uses a form of IP computer networking to interconnect devices, in a "serial" fashion. This approach is generally called "Audio Over IP" (AoIP), and greatly simplifies interconnection because, like computer networking, each device can be conveniently plugged into the system at any point on an adequately high-speed LAN, usually by a single computer networking cable (even if the device includes many inputs and outputs). Routing of signal paths is done in a "virtual" or "logical" way by a computer networking process, rather than the "literal" or "physical" routing approach used by analog and earlier (non-AoIP) digital systems. These advantages are encouraging many radio stations to convert from analog or earlier digital systems to AoIP, and most new radio facilities being built today are based on AoIP systems. More detail on AoIP for radio can be found later in this chapter.

For any type of radio facility today, however, the initial conversion from acoustical sound to electrical audio signals is still performed by analog *transducers*. Microphones at the start of the chain remain analog, as do loudspeakers and headphones at the other end. Nevertheless, in a typical modern facility, it is common to have microphone signals converted to digital immediately after leaving the microphone (or after its associated *preamplifier*) and have it remain in digital form until after it has been fed into the transmitter. An FM or AM transmitter accepting digital inputs will, of course, convert the signal back to an analog waveform for RF transmission (even in the case of so-called digital broadcasting, as explained in Chapter 7).

Air Chain

A station's *air chain* is the path that on-air program material follows from the program source to the transmitter. As a rule, the arrangement for a local radio station's on-air studio and air chain, up to the *studio-transmitter link* (STL), will look something like the simplified block diagram in Figure 8.1.

Microphones and other audio program sources feed into an audio mixing console; the output of the console will often pass through an *audio delay* unit. This device, if used, provides the station with protection from the airing of momentary undesirable content, and allows it to be removed before airing in ways that are relatively unnoticeable to the listener. This can help the station avoid listener complaints, and possible regulatory actions—possibly culminating in hefty fines—from the inadvertent airing of potentially offensive material. (For this reason, this type of device is sometimes referred to as a *profanity delay*. This also differentiates the unit from other types of audio delays that may be used in the radio facility. More on this later in the chapter.) Because of the programming complexity added by the use of profanity delay, in many cases a station will only activate the device for programming where such utterances are likely, such as live call-in shows, and bypass the unit during other programming less likely to contain any such content, so real-time material can be aired as it happens.

After exiting the delay unit, the signal may pass through some sort of *distribution system* (not shown in the figure), and proceed to the *Emergency Alert System* (EAS) equipment. This equipment interrupts the radio station's regular signal when activated by an emergency alert coming from any of several officially designated sources outside the station, and inserts the content of a message using audio and data, the latter of which is intended to be received by other stations down the "daisy-chain" line that EAS uses to distribute messages. At all other times, the station's EAS unit simply passes audio content through it without alteration. More on EAS equipment operation is found later in this chapter.

Following the EAS system, the audio signal goes through various items of *audio processing* equipment that adjust the overall sound of the station and then to the STL for sending to the transmitter.

Audience Measurement Watermarking

Also at this stage of the air chain, audience measurement *watermarking* may be added, which is a low-level digital signal inserted in an inaudible fashion using perceptual coding techniques, but which can be detected by special listening devices. Each participating station's watermark is unique (and separate watermarks can be placed on multiple feeds of the same content to separately identify a station's AM vs. FM vs. Internet simulcasts, for example.) These watermark-sensitive listening devices are small and portable, and are issued by audience research companies to a statistical sample of people in a given broadcast market, who wear them as they proceed through their day. Whenever the wearer turns on a radio, the listening device detects the watermark signal (if one is present) embedded in the audio signal of the station that the user has selected, and records in its local memory the fact that the wearer heard this station at this time. Once every day or so, the receiver downloads all its captured watermark data to a central location, which adds the data to all other wearers' downloads from that time period in that market, and thereby calculates a statistical estimate of the listening audiences for all participating stations in the market. One implementation of this type of automated audience measurement system is called the *Portable People Meter*™ (PPM, not to be confused with the identically abbreviated Peak Program Meter described in the Audio Mixing Consoles section below).

All Studios Large and Small

Some of the blocks in Figure 8.1 indicate <u>categories</u> or classes of equipment, not necessarily individual units, and the interconnections show the general arrangements for signal flow, not specific audio connections. In the following sections, we describe the main equipment in each of these blocks and how it all works together. Large radio stations and network centers have many more studios and operate at a larger scale, or include facilities other than those specifically shown here, but the types of equipment and principles of operation are essentially the same.

We will start with the audio mixing console, which is the heart of the studio system, and then discuss audio sources and monitoring, control systems, and, finally, the other major items of equipment and functions that make up a typical radio studio facility.

AUDIO MIXING CONSOLES

The audio mixing console is a device that allows several program sources to be mixed, monitored, and fed to the transmitter. An example is shown in Figure 8.2. It is often referred to as a *mixing board*, *control board*, or just a *mixer, board* or *console*. Both analog and digital mixing boards are widely used; both have similar facilities and user controls, although the way they process the audio inside is completely different.

The mixing board has multiple signals fed into it from different program sources, such as microphones, CD players, hard disk recorders, or computer automation playout system. The board has controls that allow the operator to select each source and feed one or more simultaneously to a mixed program output called a *bus* (from "bus bar," a metal junction bar joining two or more electrical circuits in traditional analog systems). Each input has a *level* (volume) control known as a *fader* (also known as a *"pot"* from the name *potentiometer*, the *variable resistor* component used in most mixing boards). In earlier times, faders were rotary controls, with clockwise motion used to turn up the signal; today,

FIG. 8.2. Audio Mixing Board. Courtesy The Telos Alliance

sliding *linear faders* are nearly universal, and the control moves in a straight line (up is louder). Other input channel controls usually include a *pan-pot* (to adjust where a mono sound source appears to be located when heard on a stereo system), *balance* for stereo channels, and, in some cases, *equalization* to adjust the sound for treble, bass, and so on.

Modern "Control Surfaces"

Many digital and all AoIP mixing consoles have very little actual audio hardware inside. Instead, they act simply as remote controllers to external audio processing equipment racks (or "engines") that can be housed in separate rooms and are connected to the console by a computer networking or proprietary cable. In these cases, all audio inputs and outputs listed below are physically connected to outboard interfaces and not to the console directly. Such distributed interconnection can further simplify design and installation of such systems.

Inputs

Board inputs are of two main types: microphone and *line* inputs. Each feeds a *channel* on the mixer (not to be confused with RF channels used for transmission). Sometimes these channels are called *input positions*. Microphones have very small signal outputs and require sensitive channel inputs to raise their signals to more manageable levels and keep them free from interference that may be picked up along the signal path. This function is performed by the microphone preamplifier mentioned above, which is nearly always analog today. Other than microphones, audio equipment and distribution systems output audio at the much higher *line level* (about 1000 times stronger than *microphone* (or *mic*) *level*, electrically speaking). Audio levels are usually quoted in *decibels* (dB), which is a statement of a ratio compared to a standard level. The reference level is usually quoted as "0 dB." This means that in most cases, a somewhat counterintuitive principle applies, by which 0 dB is not the <u>absence</u> of an audio signal but in fact the maximum or preferred operating level, with most signals existing at negative levels (e.g., -10 dB, read as "minus ten decibels") below the 0 dB reference. More on this is found in the section on Monitoring below.

Many mixing boards, whether operating as analog, digital or AoIP systems internally, are equipped to accept both analog and AES3 digital line level inputs, along with analog microphone inputs.

Outputs

Most radio consoles offer the possibility of creating multiple simultaneous outputs, each with a different mix of audio sources. In some cases the differences between mixes will be simply the presence or absence of a particular source, by virtue of that source's input being *assigned* (or not assigned) to a given output bus. In some cases, the main output is called the *Program* bus (or Program 1), and a secondary output is called the *Audition* bus (or Program 2). In other cases, the buses are simply numbered 1 through *n*.

Another approach allows completely separate <u>mixes</u> to be attained, in which the assortment of sources <u>and</u> their relative levels can be differentiated between outputs. These are called *auxiliary mixes* (or *"sends"*), and are usually achieved by adjusting rotary pots on an input position "strip" in the console for the desired level of that input into the auxiliary mix. These mixes can be made independent of the main fader's control of that input (known as a *"pre-fader"* send), or they can follow both the level adjustment of the fader as well as the rotary send pot on the input strip (a *"post-fader"* send). They are often used to feed a separate mix of program content to recording devices, allowing a station to record a version of the program that is slightly different from the on-air broadcast (perhaps to be used for a differently edited program on a later rebroadcast, for example).

Radio boards also include some specialized features not found on general audio consoles. One of these is the so-called *mix-minus* feed, which is usually the full program mix except for one input not assigned to the mix-minus bus. This bus is typically used to feed audio to callers when putting their calls on the air, thereby preventing echo to the caller, and/or feedback on the air. Mix-minus may also be used on occasions as a *foldback, return,* or *cue feed* to other live contributors (not necessarily on the phone, but typically at some remote location). More on this subject can be found in the Telephone Hybrids section below.

Monitoring

Audio monitoring is provided to allow the operator to listen to the program output or selected sources to check for program content and quality. Audio monitoring may be on high-quality loudspeakers in the control room, but, if the control room is also the studio in a combo arrangement, headphones are used to prevent interference with live microphones. Sometimes a small loudspeaker is also built into the console.

Another specialized feature of radio boards is *cue* or *prefade listen* (PFL) *monitoring,* which allows a source to be monitored before it is *faded up* for use. Often this feature is engaged by clicking the rotary or linear fader downward below its lowest volume position, and is therefore sometimes called "sub-fader cue." The cue or PFL audios usually sent to a small speaker separate from the main monitor loudspeakers, to allow the operator to clearly differentiate the audio coming from the "cued" source being readied for air from the audio that is currently on the air.

Meters are provided to give a visual indication of the program level. This allows the operator to accurately adjust the output so it will correctly *modulate* the transmitter and not be too loud or soft. Meters may be of several types, the most common being the *volume unit* (VU) *meter* and the *peak program meter* (PPM), which measure and display audio levels in slightly different ways. VU meters have been used traditionally in U.S. radio stations, but modern consoles may also use peak indicators. Either sort may be actual mechanical meters or electronic displays.

It is important to note that neither of these meters directly corresponds to the *perceived loudness* of the sound to listeners, however. VU meters and PPMs are intended to measure the electrical level of an audio signal, so that it avoids overloading a circuit with too much signal, or becomes susceptible to noise buildup from being at too low a level. But two sounds with the same VU meter or PPM readings may significantly differ in their apparently volume to a human listener. This is the difference between level and loudness, and it has led to the development of loudness meters, which attempt to present a visual indication of perceived volume by the average listener. More on this subject can be found in the Audio Processing Equipment section below.

Effects and Processing Units

Some types of productions need more than the basic equalization controls built into mixing consoles, to change the way some audio sources sound. Audio consoles used for production and recording therefore allow additional *outboard devices* to be inserted into the mixes they produce (or occasionally include them as built-in components). These devices allow more complex equalization, or provide effects such as *echo* and *reverberation*, or *pitch change* (where a voice or sound is made to sound higher or lower than it really is). Such units use *digital signal processing* (DSP) to produce the desired sound. Processing such as *dynamic range compression* (see section later in this chapter on audio processing equipment) may also be used, and some stations add a separate *microphone processor* for tailoring the sound of the presenter's microphone before feeding it to the console.

Operation

To illustrate the role of the mixing board in producing a live on-air radio program, consider, for example, the sequence of events that occurs when a radio announcer is talking between two program segments coming from two CD players. While the first segment is playing, the next CD is loaded and *cued* using the prefade listen monitor (i.e., the correct track is selected), and the device is left ready to go at the start of the desired song or program. At the end of the first segment, when the announcer wishes to talk, the first CD player is *faded down* (i.e., the fader is moved to decrease the level), and the channel switch may be turned off. Then the channel switch for the microphone input is turned on (if not already on), and the microphone is faded up, or *potted up*, and the announcer starts talking. At the same time, the channel for the next CD that will be played is potted up, although no audio is heard because the CD has not yet been started. As the announcer finishes his or her introduction, the channel switch for the CD player is turned on, the start button is pressed, and the music begins playing (the start action may in fact be automatic from the act of turning the channel on, or moving the fader, depending on the design). At this point, the microphone is potted down.

This is just one example of how a mixing board is used. There are many other scenarios using different input sources, which may be mixed simultaneously. Often a music piece is started while the announcer is still talking; this helps ensure that there is no silence, or *dead air*. When an announcer is talking over background music or other sounds it is called a *voice-over*.

MICROPHONES

Microphones (often called *mics*) convert sound waves created by human voices, instruments, or other acoustical sources into electrical signals. For this reason, they are considered *transducers*, the technical term for devices that convert one form of energy into another. How well this conversion is accomplished is a measure of the *fidelity* of the transducing process.

Although each microphone model is designed a little differently, they all have generally similar design principles. Microphones are quite straightforward and are fundamental to radio broadcasting, so we will discuss the inner workings of the two most commonly used microphone types: *moving coil* and *condenser*.

Moving Coil Microphone

In the *dynamic moving coil* (or simply *dynamic*) microphone, shown in Figure 8.3, a small drumhead-like surface called a *diaphragm* is exposed to the incoming sound pressure waves in the air, and moves backward and forward in a corresponding

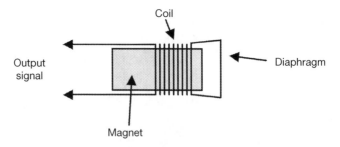

FIG. 8.3. Dynamic Moving Coil Microphone Design

manner. The back of the diaphragm is connected to a metal coil of wire that slides up and down over a small magnet when the diaphragm moves. The coil's movement through the magnetic field around the magnet causes an electrical signal to be created in the wires in the coil. This signal is a reproduction, in electrical form, of the sound waves that hit the diaphragm. The ends of the coil are connected to the plug on the end of the microphone, and the signal can be fed from there to a microphone input on mixing board.

A different type called the *dynamic ribbon* (or simply *ribbon*) microphone operates using a similar principle. In this case, a very thin piece of metal foil (the ribbon) is suspended in a magnetic field, and exposed to the sound pressure wave. When sound waves move the ribbon through the magnetic field, an analogous electrical signal is created in the ribbon itself.

Condenser Microphone

The condenser microphone, shown in Figure 8.4, operates using a different principle based on the operation of a *capacitor* (also known as a *condenser*). A capacitor is an electronic device with two conductive *plates* that allow electricity to flow from one to the other at a rate dependent on the material between them and their distance apart. In the condenser microphone, incoming sound waves strike a flexible, electrically conductive diaphragm, which is situated in front of a stationary metal back plate. Together, the diaphragm and the back plate form a capacitor. The distance between the plates varies slightly as the diaphragm moves in accordance with the sound waves it encounters. So, if electricity is applied to the capacitor in a condenser microphone, the electrical flow will vary as the capacitor's diaphragm moves, and a signal is produced that is an electronic version of the incoming sound waves.

The main advantage of the condenser microphone is that the capacitor circuitry is smaller and lighter than the magnets and coils used in dynamic microphones. For this reason, condenser microphones typically have higher fidelity, since their diaphragms can move more responsively to the sound waves they transduce into electricity. Their small size makes them appropriate for the small lapel or clip-on microphones often used in television for on-camera talent (called *lavaliere* mics). Larger high-quality condenser microphones are often

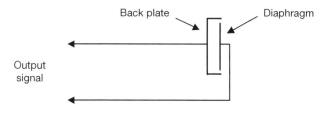

FIG. 8.4.   Condenser Microphone Design

used for studio recordings. The disadvantage of condenser microphones is that they are *active* devices, meaning that they require a power source—either from an internal battery or external power supply—whereas dynamic microphones are *passive* devices that require no power to operate. Owing also to condenser microphones' active design, their output levels are generally higher than that of dynamic microphones.

Microphone Patterns

A microphone pattern refers to the directions from which it picks up sound. Microphones used in broadcasting come in four main pattern types. The *omnidirectional* picks up sound from all directions; the *cardioid* or *unidirectional* picks up sound mainly from one direction; the *figure-of-eight* has a pattern like its name, and picks up sound from two opposite directions; and finally, the *shotgun* is a highly directional microphone.

A microphone may also be used in conjunction with a *parabolic dish* reflector (shaped like a small satellite receiving dish) to create an extremely sensitive directional microphone for picking up soft sounds from far away.

Omnidirectional mics (or "omnis") can be thought of as lanterns, while unidirectional mics act more like flashlights. Also analogously, directional mics of all types do not behave with razor-sharp edges to their patterns, and provide only an approximation of directional pickup. This is due to the fact that the same pickup pattern will not be applied to all frequencies of sound. Generally, a directional mic is more directional at higher frequencies, while at low frequencies, almost all directional mics act like omnis. Directional mics are also more sensitive to wind noise, *plosives* in speech (commonly called "p-pops") and *handling noise* when used in handheld applications.

LOUDSPEAKERS AND HEADPHONES

We will cover loudspeakers (also known as *speakers*) and headphones next because they operate in a similar, but opposite, manner to a moving coil microphone to convert electrical signals back into sound waves. Thus they are also transducers.

In a speaker, an electrical signal (of a much higher level than the one that comes out of a microphone) is fed into a metal coil located in a magnetic field created by a large magnet. As shown in Figure 8.5, this metal coil is attached to a lightweight diaphragm or *cone*, often made of a paper-like material. The changing electrical signal in the coil causes it to move back and forth in the magnetic field, thus causing the cone to move too. The cone's motion causes the air to move, creating sound waves that can be heard by the human ear. These sound waves, of course, correspond to the electrical signal that is fed to the

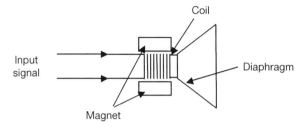

FIG. 8.5.   Typical Loudspeaker Design

speaker through the speaker wire. In this way the transducers at each end of the path convert sound waves into analogous electrical waves (called audio signals) and back again. In between, the signals in their electrical form can be stored, manipulated and transmitted over vast distances, which are the key techniques enabling the craft and business of radio broadcasting.

The size of the diaphragm used in a loudspeaker largely determines the range of frequencies it can reproduce well. Therefore, inside the box of a high-quality, "full-range" speaker system, there will always be at least two speaker units of different sizes to cover a wide audio frequency range. The speaker unit that covers the highest treble frequencies is called a *tweeter* (typically one to three inches in diameter), and the speaker that covers medium and low frequencies is called a *woofer* (usually between about 6 and 15 inches in diameter). If there is a third unit for frequencies in between, it is called a *midrange* unit. There is sometimes a special speaker in a separate box, called a *subwoofer*, to cover the extremely low bass frequencies. The component that divides the full-range audio signal into separate bands for each of the different speaker parts is called a *crossover network* (or simply a "crossover").

Every speaker system has an associated *power amplifier* that takes a line level signal from a mixing console, or other program source, and amplifies it to the high power level needed to drive loudspeaker units. In some cases, this power amplifier is built into the speaker cabinet itself, but more commonly the amplifier is a separate outboard device, connected to a speaker cabinet by a single pair of wires.

Headphones operate in a similar manner to loudspeakers, the main difference being that the electrical signal levels fed into them are not as strong, and, of course, the physical elements of a headphone are smaller than those of a loudspeaker.

No matter how many revolutionary changes occur in other broadcast equipment design, the basic operation of headphones, speakers, and microphones will probably remain basically unchanged. Although it is possible to convert all of the other audio signals in a broadcast facility to digital, headphones, speakers, and microphones have to retain their analog design because acoustical sound and human hearing will continue to operate in analog form.

CD PLAYERS

CDs, or *compact discs*, have been for many years one of the most convenient, reliable, and highest quality forms of media for long-term digital audio storage. The digital audio on the disc is uncompressed and recorded with 44.1 kHz sampling and 16 bit resolution using pulse code modulation (PCM). This provides very high-quality sound reproduction.

A CD is reliable because the information is permanently etched, or carved, into the plastic of the disc as very small indentations or *pits*. It cannot be erased by passing through a magnetic field like the information on a recording tape can, and the only way to damage the information is to physically damage the CD by breaking or severely scratching it. Small scratches are often not a problem because CD players are able to miss a few bits of digital information and still accurately reconstruct the recorded music. They can do this because the digital audio data includes additional bits, added in a specially coded manner, called *forward error correction*. These enable the CD player to determine what the correct value of a missing piece of data is. In addition, the CD format includes *error concealment*, which tries to cover up small errors that it cannot completely correct. An error correction or concealment system is only capable of fixing errors up to a certain point, however. If there are too many missing or damaged pieces of data, all attempts at recovery will fail, and the CD will skip or stop playing.

A CD is also durable because it is not subject to any wear and tear during the playback process (unlike vinyl records and magnetic tape, which require physical contact with equipment during each playback). A CD player reads information by shining a light (actually a laser) onto the disc and analyzing the reflections of this light that are caused by the pits recorded on it. Because there is no mechanical contact between the laser and the disc, there is no ongoing wear and tear.

Analog and Digital Connections

Most professional CD players have both analog and digital outputs. If the station's audio system is analog-based, then the analog outputs can be used to feed a signal into the mixing board. If, on the other hand, the station's audio system is digitally-based, then the digital outputs can be used.

Other CD Variants

Equipment is available for recording uncompressed audio and compressed audio files on compact discs. These devices are not widely used for radio studio operations, but they may be used for archiving audio content, to take advantage of the robust CD format.

HARD DISK RECORDERS AND AUDIO WORKSTATIONS

Hard Disk Recorders

Chapter 6 explained how computer hard drives can store large amounts of data as files on magnetic disks. Because digital audio is just another form of data, it can be stored in this way. Fundamentally, all such systems use a computer with some kind of interface for the audio input and output signals (analog, AES3, or AoIP). Once the audio is stored as a data file, it can be transferred to and from other digital storage devices via a computer network, either in the same studio or across the country.

Audio files on the hard disk system may be stored in uncompressed form (sometimes known as *linear* recording) or, to reduce the size of the files, audio files may be stored with some form of perceptual coding (see section below on audio data compression).

There are several different approaches to storage of audio on hard disk. The most common method today is based on standard computers, with the usual keyboard, mouse, video monitor and speakers for the operator interface. Alternatively, a dedicated, standalone piece of hardware can be used, which typically provides front-panel controls with buttons, knobs, and indicators for specific functions, along with audio input and output connectors on the back panel.

The process of capturing real-time audio and converting it to a digital audio file for hard-disk storage is generally called *ingest*, and the retrieval of the audio from hard-disk storage is generally called *playout*. In the latter process, the hard-disk recorder or associated computer reads the file from the disk, formats it into an appropriate analog, AES3, or AoIP data stream, and sends it out through a cable to an appropriate input on a mixing board, switcher or other destination.

Digital Audio Workstations and Servers

Digital audio workstations are computers with specialized software providing functions such as automatic ingest of audio material, editing and audio processing of content, file conversion, file management, as well as basic record and replay. Most workstations also include professional audio *sound cards*, which replace a computer's built-in audio inputs and outputs with high-quality audio interfaces using professional audio connectors, typically on outboard boxes that connect to the workstation via a USB port.

Digital audio workstations may be used *off-line* (i.e., in isolation), or they may be associated with a production studio or on-air studio. In all cases they will typically be connected to a computer network and hence be able to share files. A system that is used to provide central storage of files is called a *file server*. Any other authorized workstation on the network can simultaneously access any file stored on the server. The file server, therefore, provides the functions of both a library and a distribution system.

Songs, commercials, newscasts, and all other types of audio segments used in a broadcast facility can be stored on a server for later recall by whatever playback device wants to use them. They can be edited, converted to different formats, and distributed to other destinations in the station or elsewhere. In addition, the server can be used for live retransmission of a digital signal as it receives it. Perhaps the most important thing about hard disk–based storage systems is that they can be easily integrated with a computer-based automation system, as described in the program automation section later in this chapter.

Because computers generate acoustical noise—mostly from cooling fans and spinning hard disks—their use can be problematic in necessarily quiet spaces like broadcast studios and control rooms. (Servers can be particularly noisy due to their often large number of hard disks, all spinning at high rates of speed.) For this reason, the computers themselves are often housed outside the studio or control room, and special remote interfaces are used to connect them to the keyboard, video monitor, and mouse that users require for operation of the workstation from inside the studio or control room. These are called *KVM interfaces* (or "extenders"), and they are required because normally the keyboard, video monitor and mouse (KVM) are connected to a computer by cables only a few feet long, while remote location of the computer may require hundreds of feet of wiring for these connections. KVM hardware allows such longer connections. For convenience, it is also possible to have one user workstation (consisting of a single keyboard, video monitor(s) and mouse) be switched to control any one among multiple remote computers, using a *KVM switch*.

Digital audio workstations can also be run on laptop computers for portable application. In this case, the audio interface is typically housed in an outboard unit connected to the laptop via USB or a similar interconnection.

Digital Audio Editing

Audio editing is a primary feature of digital audio workstations. Program material can be selected, cut, reordered, modified, and generally prepared for transmission, while visual aids to the editor—such as a depiction of the audio waveform—are displayed on the workstation's video screen. Usually at least some of the functions of an audio mixing console are provided, allowing the sound to be modified and several sources to be mixed together. One big advantage, compared to traditional tape-based editing, is the ability to access all program segments with *random access* instantly from hard disk storage. Another advantage allows the user to quickly "undo" an edit, or go back and revise edits already made early in the program without affecting later edits. This *random access* feature allows *nonlinear editing*, which is in sharp contrast to having to shuttle a tape backward and forward to locate a particular program segment. Most workstations also employ *nondestructive editing,* which means that while edits are being composed, the original, unedited audio file is still retained in storage, and can be accessed later for full playout or repurposing with different edits.

Audio Data Compression

The digital bit rate from a CD player is more than 1.4 megabits per second. This means that long segments of music or other programs will require very large files when recorded on a hard disk.

A technique called *audio data compression,* based on *perceptual coding,* may be used to reduce the amount of data needed to transport or store digital audio (see Chapter 5). Note that this type of compression has nothing to do with the dynamic range compression that takes place in audio processing (see later sections in this chapter). There are many different audio compression systems, often referred to simply as *codecs,* and most have been standardized by some international standards organization. A specific codec is usually referred to by some abbreviation of its standard's name (such as "MP3" for the *MPEG-2 Audio Layer 3* coding standard).

The most commonly used in radio are *MPEG-1 Audio Layer 2* ("MP2"), *MPEG-2 Audio Layer 3* ("MP3"), and several variants of *MPEG-4 Advanced Audio Coding* ("AAC").

The continual increase in the storage capacity of hard disks, coupled with reductions in their cost and increases in their data-access speeds, has reduced the need for data compression in hard-disk audio storage. Therefore, most facilities now store audio on hard disk without data compression, which increases audio quality.

RADIO PROGRAM AUTOMATION

By definition, an automation system replicates the work of one or more persons, performing repetitive tasks according to preprogrammed rules and at specified times. In the case of radio broadcasting, a program automation system is a software program that runs on a computer and generates all or part of a program service's content. The basic task is to assemble program segments in the correct order and play them at the proper times. Most automation systems can control external devices, such as CD players or tape machines, using remote control contacts, but the greatest advantages come when the program material is stored as audio files on a hard disk storage system that the computer controls. Automation systems can also control a facility's audio routing switcher (via control-data interface), allowing major reconfigurations of the content flow through the facility without operator intervention.

A radio automation system can run on the same computer as a digital audio workstation, or the two processes can run on separate computers that are on the same network. In either case, it is fairly straightforward for the whole process of program creation, scheduling and audio playout to be combined together.

If live presenter announcements are not required, it is possible with a fully featured audio workstation, to produce and send out a complete program without using an audio mixing console at all. Most stations, however, need the flexibility that a traditional mixing console provides and usually use both systems together, with the digital audio workstation or automation system feeding an input of the mixing console.

Scheduling and Other Functions

Planning the program schedule and preparing the *playlist* for the on-air program may be carried out within a suite of computer applications associated with the automation system. These will be accessible from various computers on the network and will often include interfaces to other functions, such as the station's *traffic system*, which is used for managing the sale and fulfillment of advertising time, and the station's *asset management system*, which keeps track of program material and the rights to use it.

A relatively new function that is growing in importance is the management of non-audio content that is broadcast by a radio station, and often intended to accompany or enhance the station's audio content. This is generally referred to as *program-associated data* (PAD), and can include text (such as song titles and performer names), graphics (such as album cover art or advertiser logos), streaming video, or other data. Modern radio automation systems can be configured to handle such non-audio material, and when necessary, deliver it in proper synchronization with the audio content to which it corresponds.

DIGITAL RECORD/PLAYBACK DEVICES

Digital Cart

Digital cart machines provide all the functionality of a traditional analog tape cartridge machine (see below) but use internal hard disk storage to provide greatly enhanced capabilities. They have simple front-panel controls for easy recording, selection, and playback of material. Usually, they are connected to a network to access commercials, promos, or other items produced elsewhere, but most also have some built-in editing facilities. Such devices are typically used for live shows, where there is not time to schedule all the segments with the automation system.

A variant of the digital cart provides a type of control panel sometimes known as a *shot box*, with many individual buttons. In this device, selections from hundreds or thousands of individual audio segments (e.g., sound effects, or any short program segment) can be rapidly assigned to individual buttons and recalled and played instantly.

Solid-State Recorder

The advent of inexpensive, large-capacity, solid-state *flash memory* cards has enabled a relatively new generation of digital recording devices. Available in both battery-powered portable versions and fixed studio versions, these machines combine the capabilities of standalone, dedicated audio recorders with the advantage of audio files that are compatible with computers, networks and servers. These recorders have analog and digital audio input and output connections, and are also able to connect to a computer or audio workstation for file transfer. Thus, they can be used as independent record/replay devices, or as acquisition devices for a file server.

Digital Audio Tape (DAT) Recorder

An older storage format that may still be found in some radio stations is the digital audio tape (DAT or R-DAT) recorder. When introduced in the late 1980s, it offered major advantages over analog tape, because its underlying digital technology enabled it to record and play back audio with negligible noise and other degradations.

The DAT format uses 4 millimeter magnetic tape housed in a cassette. The tape head is mounted on a rotating drum (hence the R in R-DAT) and uses *helical scan* like videotape recording systems. The standard recording time is up to two hours on one tape, but a long-play mode can allow up to six hours at lower quality. The digital audio signal is uncompressed PCM, with alternative sampling rates of 32 kHz, 44.1 kHz, and 48 kHz, and 12 or 16 bit resolution.

One disadvantage of DAT compared to open-reel analog recorders was that it is impossible to edit the tape using cut and splice techniques. Also, while DAT offered greatly improved audio quality over analog tape, it was subject to some of the same mechanical problems, including malfunctions with the tape transport mechanism, and wear and tear with the tape, which could cause signal *dropout* and audio interruption or degradation. Long-term storage of DAT tapes was also a concern, and it is therefore not widely used for archival purposes.

Both fixed and portable versions of DAT recorders were offered, but production of DAT equipment ceased in the mid-2000s, and the format is currently considered obsolete.

ANALOG DEVICES

At one time, analog record turntables and audio tape recorders were the mainstay of broadcasting. To provide a historical perspective, a brief summary of the main types of such equipment is included here. Examples of these machines may still be found in most radio stations, although nowadays they are used mainly for playing archive material and transferring (*dubbing*) it to modern digital media.

Analog Turntable

Before the days of CDs, consumer music was for many years distributed on vinyl discs known simply as *records*. A 7 inch disc, rotating at 45 revolutions per minute (rpm), holds up to about 5 minutes of audio per side. A 12 inch *long-playing record*, rotating at 33 1/3 rpm, holds up to about 30 minutes of audio per side. Both sides of these discs were typically filled in commercial releases.

On a record, the audio signal is carried mechanically as variations in the width and height of a very fine, continuous spiral groove pressed into the surface of the disc. The groove begins at the outer edge of the disc and proceeds concentrically toward the center of the disc, ending a few inches away from the actual center of the disc, where the circular space remaining is typically filled with a paper label containing information about the disc's content. (Note that the CD format works in reverse, with the content starting at the innermost diameter of the disc and spiraling outward.)

Although mono disc recording began in the days of only monaural sound, stereo capability was later added. The two channels of a stereo recording are carried on opposite sides of the same groove.

To play a vinyl record, it is placed on a rotating turntable. A diamond *stylus*, underneath a *pickup cartridge* attached to the end of a pivoted *tone arm*, is placed in the groove, and the variations in the groove cause the stylus to vibrate as the disk rotates. The pickup cartridge turns the vibrations of the stylus into left and right electrical audio signals (usually using magnetic detection, like a moving coil microphone). It is therefore another form of transducer.

Cart Machine

Cart (short for "cartridge") machines are magnetic tape recorders that use plastic cartridges with an endless loop of 1/4 inch tape inside, usually running at a speed of 3.75 inches per second (ips). Carts were commonly used for playing short program segments, such as commercials, fillers, and so on. Besides eliminating the need to load and thread up an open reel of tape on a reel-to-reel tape recorder (see below), cart machines' advantages were simple push-button controls, nearly instant start, and the use of *cue tones*. These inaudible tones identified the beginning and end of a recording, so the machine could stop and re-cue itself to the beginning of the recording. This allowed the operator to simply start the machine when the content was needed, and the machine would automatically ready the cartridge for its next use. Once the cart re-cued itself, another cartridge could be easily and quickly loaded into the machine. Typically two or more machines were available in each studio (in some cases a single unit housed multiple cart "decks"), so an operator could prepare and play a sequence of short announcements back to back. Cue tones could also be used for one cart machine to automatically trigger another to start, for a semi-automated sequence or salvo, such as might be used during a commercial break.

Reel-to-Reel

Reel-to-reel recorders (also called *open-reel*), because of their long lengths of easily accessible 1/4 inch recording tape, were most useful for recording and playing back long programs, and for editing program material by physically cutting and splicing the tape. The system records stereo channels using two parallel tracks on the tape. Various tape speeds were used, the highest normally being 15 ips, which provided high audio quality. Long-play versions, using very low tape speeds, were often used for logging and archive purposes. All standard tape speeds were derivatives of 15 ips, with 7.5 ips being most common in radio for reel-to-reel recorders. Versions of this type of recorder, with tape up to 2 inches wide, were also used for multitrack music recording, although not often found in radio studios.

Cassette

Cassette recorders use 1/8 inch tape inside a plastic cassette, running at 1.875 ips, and with a bidirectional track format providing two-sided operation. Because of their compact size, these machines were useful for broadcasters in recording audio in the field, such as news interviews, and they may still occasionally be used for that purpose. Fixed versions were used to play the cassettes back in the studio, and are occasionally still found in control rooms to accommodate archival material that be found in this once popular format.

Like cart machines, cassettes suffered from the fact that tape-to-head contact was largely influenced by the shell of the plastic case, which could warp, shift or otherwise provide inconsistent performance. This was a primary reason why reel-to-reel recorders could provide better audio quality (along with the higher tape speeds they used).

Noise Reduction

One of the drawbacks of analog audio tape is the well-known *tape hiss* noise that all analog tape recordings have to some extent. High-quality open-reel machines running at higher tape speeds, and with sophisticated tape coating materials, were able to reduce this noise considerably, but slower-speed machines, particularly cassette recorders, suffered from it quite badly. The difference between the audio signal level and the hiss, or other noise on the tape, is called the *signal-to-noise ratio*. The larger the signal-to-noise ratio, the better the recording sounds. A low S/N ratio will result in a recording that sounds "hissy."

To help overcome the tape noise problem, a company called Dolby Laboratories developed a system of *noise reduction* for analog tape. Dolby® noise-reduction technology is a sophisticated form of *dynamic equalization*. During recording, the professional *Dolby A* system uses amplification of low-level sounds in many narrow frequency ranges over the entire audio range. During playback, the same sounds are suppressed back to their original level, which simultaneously reduces the noise introduced by the tape. Dolby B circuitry is a less complex—and therefore less expensive—version, which was widely used in the cassette tape format described above—both for commercial music sold in the format, and for consumers' own recordings on blank tape. It operates only at higher frequencies, where tape hiss is most noticeable. Dolby's noise reduction and other audio processing is also widely adopted in the motion picture industry and used on many film sound tracks.

TELEPHONE HYBRIDS

The interface to an analog telephone line to be used on-air requires a device called a *telephone hybrid*. This piece of equipment converts incoming audio from the phone line into a line level signal that can be fed into a mixing board. It also converts a line level signal coming out of a mixing board into an audio signal that can be fed over the phone line back to the caller.

The hybrid allows the caller to be heard on air, and the show host to hear the caller without having to pick up a telephone handset. At the same time, it allows the caller to hear the talk show host and the rest of the program. It ensures that only the caller's voice is of telephone quality, while the show host's voice remains of broadcast quality.

As described above (under "Outputs" in the Audio Mixing Consoles section), the output from the console providing the host and program audio to the caller is generally of the *mix minus* variety, meaning that the feed does not include the caller's voice, but does include all other elements of the program. This ensures that the output of the telephone hybrid is not fed back into its own input, which could cause feedback and/or echo.

When there is a need for multiple sequential or simultaneous callers to be placed on-air, multiple telephone hybrids can be used. This can be achieved by installing multiple individual telephone hybrid units in the studio, or by choosing one of many multiline, multi-hybrid systems, which are designed specifically for talk-radio applications. It is not uncommon for a talk-radio studio to have four or more telephone hybrids in a single studio, each with switchable access to a large number of incoming phone lines to the station.

REMOTE SOURCES

Most stations receive at least some content—news, weather, sportscasts, or network programming, for example—from remote sources. These sources will appear as inputs on the audio mixing console, either directly or via an *audio routing switcher* (see below). These remote contributions are often distributed by *satellite link* if they are coming from a network or organization distributing to multiple stations. A local station, receiving remote broadcasts and news reports from the field, will usually use either *remote pickup units* (RPUs—see Chapters 10 and 11) or some sort of telephone lines as links. The *terminal equipment* for such links is usually in the master control room, and the audio is fed from there to the mixing console. In some cases, streaming over the public Internet may be used when no other method is available, in which case the terminal equipment may be a computer. Chapter 11 covers the topic of remote contribution links, often called *backhaul*, in more detail.

AUDIO DELAY UNITS

The *audio delay unit* (also known as the *profanity delay*) follows the audio console and usually provides between five and ten seconds of audio storage and delay. It is used for live programming, where it is possible that a contributor (perhaps a telephone caller) may say something unacceptable on the air. If that happens, the operator can press a button that "dumps" the offending words before they leave the delay unit heading for the transmitter.

Early delay units provided a fixed amount of delay and were therefor cumbersome to operate, particularly for getting in and out of the programs that used them (if the station didn't operate with delay all the time), or after a caller was "dumped." Modern units are more flexible, however, and provide much easier operation given that they offer variable amounts of delay, and the delay can be "grown" or shrunk" dynamically during the program as needed. This would allow a live show to begin and end in real time, but be delayed by several seconds during the middle of the program when calls are taken. It also allows delay to be quickly recovered after dumping a caller, so it is once again available should further profanity be encountered later in the program.

Some audio delay units are equipped with large memory buffers, allowing a live incoming feed to be "paused" for up to several minutes (allowing a station to insert other material, such as advertising, or an emergency announcement, for example), after which the incoming program can be rejoined from where it was paused, without missing any of its content. (This also allows a station to accommodate a live incoming program following another program that ran overtime, without missing the start of the second program.) Although the content coming from the delay unit would then be behind the live, real time incoming feed, the delay unit can be set to slightly (and imperceptibly) speed up the playback from its memory, and thereby "catch up" with real time over the course of a few minutes, ultimately getting back to broadcasting the live incoming feed as it is being received.

Delay units in use today use digital circuitry internally to perform this sort of time manipulation, although their audio input and output interfaces may be either analog or digital.

EMERGENCY ALERT SYSTEM

The Emergency Alert System (EAS) is the communications network designed by the U.S. federal government to allow the President to speak to the entire nation in the event of a national emergency. It also provides state and local officials with a means of alerting their communities about local emergencies such as severe weather, chemical leaks, fires, and so on.

From an engineering standpoint, the way EAS operates is relatively simple. As shown in Figure 8.6, the EAS encoder/decoder is installed in a station's air chain so it can interrupt the flow of normal programming to the transmitter, in order to insert an emergency message.

The EAS decoder constantly monitors the transmissions from at least the two input sources to which it has been assigned, for a special digital code signifying an alert of a particular type. The two sources are typically other broadcast stations, but many stations also monitor National Oceanic and Atmospheric

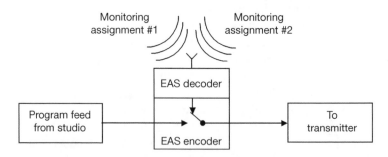

FIG. 8.6. EAS Equipment in a Radio Station Air Chain

Administration (NOAA) weather service broadcasts. Having at least two sources helps ensure that the decoder will still receive an alert message if one of its monitoring assignments happens to be off the air.

If an alert is received by the EAS decoder, of a type that the rules for the station determine should be passed on, then the EAS encoder will break into the station's air chain and transmit the alert in place of the regular programming. Encoders can be programmed to do this automatically. Alternatively, they may be set up to automatically record the alert message for later broadcast and require a person to manually interrupt the station's programming for the EAS message.

A recent addition to EAS (and a requirement for all EAS encoders at radio stations) is the support of a format for presenting emergency announcement data that can be decoded by a variety of downstream devices called the *Common Alerting Protocol* (CAP). It is an XML-based language intended for uniform use in the dissemination of alerting messages over a wide variety of delivery technologies (e.g., Internet, broadcast, text messages, etc.). While at this writing CAP is used over EAS primarily as a container to send data to first responders, government agencies and other professional communities, it is envisioned that the format ultimately could be used for delivery of detailed emergency data (including links to streaming audio and video) directly to the broadcast audience, for presentation on a wide range of typical consumer devices.

Testing of the EAS national distribution is conducted weekly, and also requires over-the-air transmissions with test announcements once per month.

AUDIO PROCESSING EQUIPMENT

Reasons for Processing

One purpose of audio processing equipment is to create a "signature sound" for the radio station, or at least to take the audio that comes from the microphone, CD player, or other source and enhance it to make it sound better when heard on a radio. The other purpose is to prevent signals that are too high in amplitude from being sent to the transmitter. Some stations employ several different pieces of equipment to carry out a lot of processing if they are looking for a particular "sound". Other stations do less processing and might only have a single audio processor. Processing requirements for AM stations are very different from FM due to the restricted audio bandwidth and higher noise level of AM stations. IBOC channels, with their different characteristics, have another set of processing requirements.

From an engineering standpoint, the main purpose of audio processing is to maintain the level of energy in the station's audio to within a specified range, and to prevent *overmodulation* of the transmitter. Usually, the processing is done in discrete audio frequency bands throughout the range of the audio signal. One way to help understand how this process works is to imagine an *equalizer*, similar to one you might have on your home stereo or car radio. An equalizer is designed to amplify, or suppress, the signal level within particular portions of the audio frequency range. Increasing the level of higher-frequency signals will make the audio have more treble, and increasing the level of lower-frequency signals will make the audio have more bass. What is different with broadcast processing is that the amount of equalization performed is <u>dynamic</u> (i.e., it changes with time), and it is performed independently in multiple bands that are set in the broadcast audio processing device. This combination of equalization and compression provides each broadcaster with a wide-ranging palette by which it can tailor the quality of its sound, and thereby differentiate itself from other stations.

Loudness

Let's consider an example of this process for adjusting the program sound. We will assume that the processor has three different frequency bands: low (bass), midrange, and high (treble). Let's say that the station using this equipment wants the on-air signal to have as high a level (volume) as possible in all three bands. In this situation, the processor will be set to increase the signal level in each band. If a station, perhaps, wants a more bass-heavy sound, it would increase the low-frequency band, or for good speech clarity, increase the midrange band.

In a home stereo system, increasing the signal level is very simple: the level (volume) control across the whole frequency band is simply turned up. In a broadcast audio processing system, however, things are a bit more complicated, largely because FCC rules limit the *modulation*, and therefore the level (volume), of the transmitted audio.

The level of the transmitted audio is very important. The primary reason why most radio stations use audio processing is to increase the *loudness* of their signals. Loudness is not quite the same as volume or signal level; it is a measure of the <u>perceived</u> intensity of a sound. Loudness takes account of the frequency makeup of the audio signal and also its *dynamic range* (the range of levels from small to large). Thus audio signals of approximately equal audio <u>level</u> can each have very different <u>loudness</u>. Many broadcasters believe that a signal that sounds louder will be perceived by the listener as being stronger and, therefore, better. It will usually be more audible in environments with background noise, such as in a car. Original recordings and other program material often have a wide dynamic range (i.e., some parts of the program have much higher amplitude than other parts). The secret to making a broadcast station sound loud is to increase the level of the softer portions of the program material, while decreasing the level of the louder portions (i.e., to compress the audio). The aim is to keep the output level sent to the transmitter as constant as possible. This enables the station to remain in compliance with the FCC's modulation limits for the transmitted signal.

Maintaining Loudness and Controlling Modulation

The FCC sets a maximum limit on modulation for two main reasons: (1) to help ensure that broadcasters' signals do not interfere with one another, and (2) to maintain a reasonably similar audio level from all stations, providing a stable listening environment for the audience. Modulation increases and decreases with the level of a station's program material. The stronger the audio level of program material is when it is fed into the transmitter's exciter, the greater the modulation level.

Several pieces of equipment are typically used to make a radio station's signal sound as loud as possible, while maintaining modulation within permitted limits: equalizers, compressors/expanders, limiters, and clippers. They are generally installed in a station's air chain in the order of the boxes shown in Figure 8.7.

Although shown as separate items of equipment in the figure, the equalization and compression/expansion functions are often performed by the same device. Equalization is needed to perform the boosting or suppression of the signal level over the appropriate frequency ranges. Compression increases loudness while expansion ensures that low-level (quiet) signals, such as background noise and electronic hiss, are suppressed and not amplified to the point that they become annoying. (As noted above, the compression, expansion and equalization processes are often combined to provide dynamic equalization in three or more independent bands.) Limiting is needed to suppress any peaks in the signal that exceed the FCC modulation limit after compression, and clipping "chops off" any excessive peaks that make it out of the limiter.

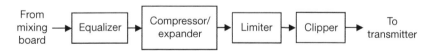

FIG. 8.7.   Audio Processing Equipment in a Typical Air Chain

SIGNAL DISTRIBUTION

Audio Interconnections

The way audio signals are transported between different devices and areas in a radio facility can be performed in either analog or digital means. Analog signals use a type of cabling called *twisted pair* (two wires twisted together), which may run in individual cables, or in large *multicore* cables with many separate twisted pairs bundled into a large cable carrying many independent signals. Such cables frequently have a metal *shield* wrapped around the inner wires to keep out interference. AES3 digital audio signals may use similar twisted pair cabling, and these can carry both signals of a stereo program on a single twisted pair, or on *coaxial cable* (with a single center conductor surrounded by an outer conductor that acts both as a shield and the second conductor for the program signal). AoIP signals are carried on *CAT-5* or similar cabling, as used for computer data networks. (CAT-5 refers to an ANSI standard defining the performance capabilities of cables with four twisted pairs of wires.) The serial nature of AoIP allows a single CAT-5 or similar cable to carry a large number of independent audio signals simultaneously. For very long runs, either within a large facility or between facilities (over distances of many miles), AoIP signals may be carried over fiber-optic cables. Within major production centers, a large AoIP system may use both types of cabling, with fiber-optic on the paths between production rooms or zones of the facility, and CAT-5 wiring within each room.

Patch Panel

To allow signals to be checked or re-routed at different points in the distribution system, cabling is often brought to a *patch panel*. This is an array of special sockets where default signal paths may be interrupted and reconfigured, using a *patch cord* (a short cable with a patch plug on each end).

Routing Switcher

Larger stations will often use an electronic *audio routing switcher* (also known as a *router, switch* or *matrix*) to supplement or replace the patch panels. This unit takes inputs from multiple sources and allows each one to be connected to one or more destinations. Each output may be connected to any one of the inputs at any given time. The component in the switcher that allows an input to be connected to an output is called a *crosspoint*. Control of the switcher may be from a local panel (usually in the master control room) or there may be one or more remote control panels at particular destinations. Such routers may be of virtually any size, from a few sources and destinations to many hundreds on inputs and outputs. Figure 8.8 shows the general arrangement. Different types of routing switchers are used for analog and digital signals. AoIP systems allow such interconnections to be made virtually, by assigning appropriate addresses to audio packets, in IP *unicast* (one output to a single input), *multicast* (one output to multiple inputs) or *broadcast* (one output to all inputs on the system) modes.

One important feature of a routing switcher is its ability to route many different signals to a particular place—such as an input channel on a mixing

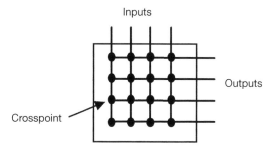

FIG. 8.8. Routing Switcher Configuration

board—through only one connection or command. As an example, a station may have many remote sources that it uses at different times, but a particular studio may only need to use one or two at any given time. So, instead of wiring all of the remote sources directly to the studio mixing board (which would take up numerous input channels there), they are connected as inputs on the routing switcher. In this case, perhaps two routing switcher outputs are connected to the studio mixing board inputs. Then, by using the routing switcher control panel to select the source required, the operator has access to all of the remote sources he or she may require, but taking up only two of the mixing board inputs. AoIP systems allow even more flexibility, with each input on a mixing console being theoretically able to connect to every available output of the system. Generally, however, an AoIP system is initialized with set-up programming that provides default access to the most commonly required sources of a given input, although this can always be changed with a few keystrokes in the configuration application of the system.

Distribution Amplifier

When routing switchers are not involved, it may be required to send one audio source to many destinations simultaneously. In that case, a *distribution amplifier* (often referred to as a "DA") is used for analog and AES3 systems. This small piece of equipment takes one input and produces many outputs of the same signal. It has two other features: One is that the output signals are isolated from each other, so a fault (perhaps a short) on a cable going to one destination will not affect the signal to another destination. The other is that the level of the signal (for all outputs, or for individual outputs independently) can be raised if necessary to compensate for losses in a long cable run. Different types of DAs are used for analog and digital signals, and either type may incorporate a *cable equalizer*, which adjusts the frequency response of the amplifier to compensate for the high-frequency loss of the cable. AoIP systems generally do not require distribution amplifiers.

IP-BASED STUDIO INFRASTRUCTURE ("AUDIO OVER IP")

As mentioned above, Audio over IP (AoIP) systems use a computer networking approach to interconnect audio streams. This method allows the economies of scale generated by the vast computer networking industry to be applied to the needs of audio routing and switching required by radio broadcasting. Instead of the "parallel" crosspoint switching methods traditionally used by analog or early digital audio systems, AoIP offers a "serial" approach to audio routing, which allows the use of far less cabling and less costly, purpose-built switchers, since off-the-shelf Ethernet or IP routers can be used for AoIP systems, with specialized radio equipment added. This approach also offers relatively quick and simple installation. For all of these reasons, radio facilities are moving toward AoIP architectures for their facility infrastructure and routing designs.

ANCILLARY SYSTEMS

Clocks and Timers

Radio stations must have accurate time for scheduling and live announcements, and time displays in all rooms in the radio facility must be coordinated. Therefore radio facilities usually install a master clock system, which provides multiple slave clocks with analog faces and second hands, or digital displays, in all studios and control rooms. The master clock is a highly accurate device that is usually synchronized to an external reference such as *global positioning system* (GPS) signals received from satellites, or other standard time updates received over the air or via a phone line from various time-reference sources. The master clock also provides accurate time to the program automation system and other computer systems in the facility that need it.

Digital *timers* are used by operators, producers, and announcers to accurately time program segments. They often include a countdown feature, which indicates how much time remains until the end of a program segment, or before the next event is scheduled to begin.

Intercom and Talkback

Communication is frequently required between operators and talent in different control rooms and studios. An *intercom* system usually covers all technical areas in the facility and allows anyone to talk to anyone else. A *talkback* system usually works between specific areas (e.g., a control room and studio). For instance, it allows a control room operator or producer to talk to talent in the studio. Where talkback and intercom have loudspeakers in a studio with microphones that may be used on-air, the loudspeaker is automatically *muted* whenever a microphone is in use and *live* (or *open*). Talkback to on-air talent often feeds their headphones, so communication can continue even while microphones are live. In some cases, radio talkback is set to feed only one ear of talents' headphones, making it easy for talent to distinguish talkback audio from that of the on-air program.

On-Air Lights and Cue Lights

On-air lights warn people in the vicinity of or inside a studio that a broadcast is in progress. These are also called *tally lights*. Depending on the system design, the lights may be on whenever the studio is in use or they may be turned on only when a microphone in the studio is turned on and open to the air.

Cue lights are sometimes used to signal to talent in a studio when to start or stop speaking. They also can provide visual indication to talent that their microphone is live to air. The controls for these lights are in the control room, and the lights can be turned on or flashed by an operator or producer. (In other cases, when talent can see control room staff through a studio window, hand signals are used for cueing talent.)

RADIO MASTER CONTROL

Somewhere in a radio facility will be the *master control*. This is the central technical control point where the equipment associated with program feeds in and out of the facility is monitored and managed, and the station's signal is monitored as received over the air (*air monitor*). Equipment for the EAS, audio processing, routing, distribution, and other ancillary functions, and the *transmitter remote controls*, will also usually be located here. In a small station, master control may be part of the on-air studio, but otherwise it will be a separate technical area. In some cases, a common master control will be used for multiple stations in a shared facility.

FACILITIES FOR IBOC OPERATIONS

Studio systems, whether analog or digital, that produce high-quality stereo programming for an analog service are usually acceptable for digital IBOC (HD Radio) transmission, as well. One special consideration, however, is that IBOC relies on relatively heavy perceptual coding (data compression) called HDC™, which is applied to the program audio signal as it is processed at the transmitter site. This requires some care to be taken with how the audio is handled at the studio.

One consideration is that some hard disk recording systems, portable digital recorders, and contribution or other links also use perceptual coding to save recording space or transmission bandwidth. While such perceptual coding may not produce audible *artifacts* (distortions of various sorts) when used in isolation, when the audio that has been processed by such perceptual coding in storage or transmission is then broadcast via IBOC, the additional generation of perceptual coding applied by HDC may create artifacts that become noticeable. To avoid this, perceptual coding used on audio content prior to broadcast via IBOC should be kept to a minimum.

IBOC processing also uses a 44.1 kHz sampling frequency (see the digital audio section in Chapter 6), as is standard for compact discs. On the other hand, most professional digital studio equipment uses 48 kHz or higher sampling. For maximum compatibility, studios for IBOC should be designed to use 44.1 kHz sampling throughout, or 48 kHz throughout with conversion to 44.1 kHz being done once at the transmitter. Failing that, the number of conversions between different sampling rates should be kept to a minimum.

IBOC stations may also wish to apply different audio processing on the analog and digital versions of their main program service, so studio routing and processing chains should be set up to accommodate such parallel processing paths.

Finally, note that IBOC broadcasting typically introduces a fairly lengthy delay (on the order of five seconds) to the broadcast signal, making air monitoring during live programming difficult. (See Chapter 13 for more detail.) For this reason, station monitoring should provide options for monitoring around the facility, such that monitoring of the broadcast output signal can be easily switched from air monitor (i.e., the received signal over the air) to *program monitor* (i.e., the outgoing signal to the transmitter prior to broadcast).

Multicasting

FM stations implementing IBOC transmission can elect to broadcast multiple separate programs within a single FM channel. This process is called *IBOC multicasting* (or *HD Radio multicasting*). (See Chapter 13 for more on IBOC multicasting.) Naturally, stations choosing to present multicast services must have studio facilities with the capacity to originate, route and monitor multiple simultaneous programs.

RADIO DATA SERVICES

FM radio stations have multiple options for transmitting data alongside their audio programs, sometimes called *metadata*. Some such data is *service-related*, meaning that it describes parameters about the station itself, such as its call letters, nickname (e.g., "B104"), or format. Other such data may be *program-related*, such as the title and artist of the currently playing song, or the current program name. Basic data broadcast systems present such information in textual form, while more advanced systems allow graphical content to also be provided.

Stations choosing to broadcast such metadata require the facilities to generate it, and in the case of program-related metadata, to present it at the proper time so that it is synchronized with the audio content to which it corresponds. This data is usually generated on a computer at the station, and fed to the appropriate encoder in the air chain, which inserts it into the broadcast signal. In some cases, the source of the data is a remote location (e.g., a database of CD cover graphics or advertiser logos), so the station's data computer must manage the feed to the *datacasting* encoder appropriately.

A variant on this technique is called hybrid radio, in which a radio receiver that also has an Internet connection (such as a smartphone with a radio tuner) is instructed by the broadcast data to access service- or program-related content that is provided to the device via its Internet connection rather than via the over-the-air tuner. Here again, a computer application manages the metadata system, either at the station, or at a central management location, and keeps the services coordinated. Hybrid radio services may also include *backchannel* features, by which the listener can use the communication services available on the receiver (e.g., the phone or Internet browser on the radio-equipped smartphone) to respond to prompts or links provided in the data stream. A station may require appropriate systems to respond to or fulfill such requests, or a remote central management service may handle this process. (See Chapters 12 and 13 for more detail on radio data broadcasting.)

INTERNET RADIO OPERATIONS

When a broadcaster wishes to send content to the Internet, the audio bitstream is sent from a studio or program automation system to an Internet streaming encoder at the station, which is then sent to a *web-hosting server* at the station, or more commonly to a third party—an Internet service provider (ISP) or Content Delivery Network (CDN)—for delivery of that stream to listeners via the Internet. In many cases, this stream is a *simulcast* of one of the station's over-the-air broadcast program services, or it may be an independent online service designed only for Internet delivery. Often an Internet simulcast may use most but not all of the over-the-air program's content—for example, replacement of certain advertisements or other announcements in the air signal with alternative announcements on the Internet stream. (For more on Internet radio, see Chapter 14.)

OTHER CONSIDERATIONS

Multitrack and Multichannel Systems

Most radio stations work with two-channel stereo sound. In contrast, studios used for music recording typically have *multitrack* capabilities. This allows individual parts of a performance to be recorded separately, with 16, 24, or more separate tracks on the recorder, and a corresponding number of channels on the mixing console. This makes it easier to control and adjust the final *mixdown* to a *master* recording at a later time. Some radio stations may have similar multitrack facilities for large productions, complex remote broadcasts, and in music performance studios.

In the past, multitrack tape recording required much larger and more complex (and correspondingly more expensive) hardware compared to stereo recording, but in current, computer-based recording systems, the difference is far less onerous. Therefore many radio stations may use "multitrack" applications on the computers in their production studios for creation of advertisements, station promotion announcements (*"promos"*) and other content.

Live and Recorded Remote Events

Many radio stations present remote events, which can vary from a presenter on site with a single microphone to a complex presentation of a multiple-act concert or festival. Specialized equipment is required for all such events, to make the production of the event as simple, portable and repeatable as possible. In the case of a large-scale event, some elements can be assigned to third parties, such as a remote recording company that provides an audio truck for mixing stage performances to stereo, which is then provided to the radio station's remote crew for integration into the program.

When such an event is recorded for later broadcast, the production equipment is fairly straightforward. In contrast, when the remote event is broadcast live, additional equipment is required for backhaul of the content to the station, for monitoring of the event on location, and for proper direction and cueing of talent and support staff during the event. See Chapters 10 and 11 for more detail on remote operations.

CHAPTER 9

Television Studios and Playout Facilities

This chapter explains the facilities used for making television programs, from the cameras and microphones in the studio to the point at which the program leaves the studio on its way to the transmitter. We will cover the major parts of the production and distribution chain, and then discuss particular studio systems and equipment.

STATION AND NETWORK OPERATIONS

In the United States, each local broadcast TV channel comes from a station with a transmitter in or near the community of license, almost always with a studio facility somewhere within the market area. Although some programming (e.g., local news) is usually produced in studios at the station, many programs actually come from external sources, such as the programming network(s) with which the station is affiliated, or from various *program syndicators*. While syndicated content from the latter entities is typically distributed as recorded programs sent to the station in advance, network content (news, sports, and entertainment) is generally distributed live to affiliated stations around the country from a *network release center*, via a *network distribution* system. Figure 9.1 is a typical, but much-simplified, television network block diagram showing the end-to-end signal flow.

As indicated in the figure, some network programs may be produced in studios at the network center, but many come from *production* and *postproduction* centers elsewhere. Delivery of finished programs to the network may be via an audio/video feed, as a recorded program on tape, or as an electronic file. A similar, separate distribution arrangement may also be in place for commercials that are to be aired on local stations.

This basic description should provide you with an idea of the complex ecosystem by which television content arrives at every television station prior to broadcast. The single stream of content sent to the station's transmitter thus

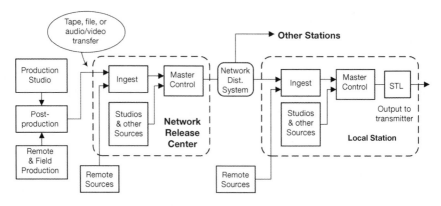

FIG. 9.1. Television Network Block Diagram

results from convergence of many separately contributed "tributaries" of content, which are properly prepared and assembled for broadcast in real time at each station's studios.

For any of this content, the postproduction process (in which recorded material is edited and assembled into finished programs) may take place at the same location as the production studio (where content is initially recorded), or it may be performed in a separate, specialized *post-house*. Network centers and local stations also have editing and postproduction systems for news and other programming.

Studio and postproduction equipment and systems are usually similar, regardless of whether they are at post-houses, network centers or local stations (although levels of sophistication and cost may vary somewhat). Master control playout facilities at the network center and local stations also have many features in common. The major items of equipment and functionality for all these systems are described in the sections that follow, while arrangements for remote broadcasting from outside the studio are covered in Chapter 10, with program links described in Chapter 11.

Centralcasting

In the *centralcasting* arrangement, station group owners may produce or originate programming (particularly news) at a central location and distribute it to their individual stations at remote locations. The operation may go further than this, and for some or all of the day, a local station's entire program schedule may originate at the central location and be distributed to the station only for transmission. This potentially reduces the amount of equipment and staff needed at each station, at the expense of possibly reducing the local flavor of some station programming.

TYPES OF STUDIOS

As with radio, television studios are special rooms or spaces in which programs are produced for broadcasting, but in this case, for both sound and pictures. Studios at local stations, network centers, or production centers vary considerably in size and overall capabilities, but most are based on the same principles.

Studios may range in size from small rooms, with one or two cameras for simple presentations, to large production studios of 10,000 square feet or more, with many cameras and space for complex shows. The larger studios are usually associated with production centers, where they are sometimes referred to as *sound stages*, from the term used in the film industry for a studio where movies with sound are shot. A local television station usually has a midsize studio for live news and weather presentations, and may also have other studios for recording special productions or commercials. Often, one studio may be arranged with several sets in different areas, which can be used for different parts of a program, or even for different programs.

Unlike radio, the on-air control point for television broadcasting, whether at a network release center or a local station, is not in a studio but is found in a separate *master control room*.

STUDIO CHARACTERISTICS

A television studio obviously has to be large enough to accommodate all the talent and the *set* (or sets)—the area where the talent's action takes place, which typically includes furniture and a *backdrop*. The studio also needs enough space for the number of cameras used to shoot the action, with space to move around the studio floor, along with other technical equipment such as *boom* microphones (highly directional microphones mounted on poles that can be held over talent to pick up their voices without the microphone appearing in the camera shot). In some cases, large studios may even include a seating area for a studio audience.

As with radio studios, television studios are usually built to be isolated from outside sounds, and they are fitted with *acoustic treatment* to control the amount of *reverberation* of sound produced inside the studio. Soundproof doors provide access for talent and production staff, often using two sets of doors with a small room or space between them called a *sound lock*. In addition, very large soundproof scenery doors are typically provided to allow scenery and sets to be brought in and out from adjacent storage and construction areas. Except for simple studios with fixed cameras, the floor has to be hard and very flat, to allow cameras mounted on *pedestals* to be moved smoothly.

Television studios have one or more associated control rooms, where the technical operators, production staff, and some of the technical equipment are located. These rooms are also acoustically isolated from exterior sounds and from one another, and acoustically treated for control of internal reverberation. In many facilities, at least one control room is immediately adjacent to an associated studio, with a soundproof window between the studio and the control room to provide production staff with a direct view of the action in the studio. In other cases, control room staff only sees the studio through the cameras, with no direct view into the studio.

So that the studio lights stay out of view of the cameras, and have adequate distance from the set to provide even coverage, they have to be hung at a high level. The *lighting grid*, from which the light fittings are suspended, are typically 30 feet or more above the floor in a large studio, so high ceilings are necessary. In addition, a track is usually provided around the edge of the studio, just below the ceiling, from which large curtains can be hung as backdrops for sets. This is the *cyclorama* (sometimes called the "cyc," pronounced "sike"). Often a light-colored background is used (either a curtain or a hard surface), which is then lit from below or above with special lighting (*cyclorama* or *cyc lights*), so decorative colored backgrounds can be produced by selecting suitable light settings.

Frequently, the studio has extra height above the lighting grid to accommodate large air-conditioning ducts, which bring in cold air and remove the heat produced by the studio lighting. They have to be large so that the air passing through the vents can be kept at low velocity, thus avoiding the sound of rushing air being picked up by microphones. This type of low-noise air-conditioning is also needed in radio studios, but is much easier to achieve there, because the lighting heat loads and studio air volumes are far lower and less cold air is needed.

Studio Lighting

It is necessary to illuminate sets and talent with sufficient light for the cameras to produce good pictures. The light must be at a consistent *color temperature*—it is bad to mix lights that have a high color temperature (tending toward blue, like midday daylight or some types of fluorescent lights) with those that have a low color temperature (tending toward yellow or red, like a regular household light bulb, or daylight at sunset). Lights also need to be *focused* in the right places and directions to achieve the desired artistic effects.

Light Fixtures and Fittings

Special lighting fixtures, correctly known as *luminaires*, are suspended from an overhead lighting grid, often with *telescopic hangers, pantographs*, or *lighting hoists* or *battens*. Some may be mounted on floor-standing tripods. *Soft lights* are used for diffused lighting to reduce shadows, and *spotlights* of various types provide lighting with more control of direction, and harder shadows. Spotlights typically have a glass lens in front of the light source to direct the light in the desired fashion. Cyclorama lights and *ground rows* are used to evenly illuminate areas of backdrop. In front of their lenses, light fixtures can be fitted with colored filters, known as *gels*, to add color. Special fabric called *scrims* can be used to diffuse light. Spotlights can also have *barn doors* and *flags*, which are movable metal shields on the front of the light fixture that can be adjusted to restrict where the light falls.

The actual light source (bulbs or tubes) used in these fixtures is one of several different kinds, both *incandescent* (with a filament) and *discharge* (without a filament), with different color temperatures and various advantages and disadvantages for each. Power may range from a few hundred watts up to 10 kilowatts (kW) for a single lamp. Most studio lights are fed from *dimmers* controlled from a *lighting console*, which is usually in the *vision control* room or area. More recently, LED lighting has been introduced and is useful in some less critical lighting cases, or where light weight (e.g., for high portability) or lower heat loads are required.

Studio Control Rooms

In a typical television facility, one or more studio control rooms house staff and equipment for the following:

- Lighting control
- Video (camera) control
- Audio control
- Production control

Depending on the overall design and space available, these may be separate control rooms or they may be combined together in one area. It is common for the sound control room to be separate, so that loud sound levels do not distract other production staff. Depending on the program's requirements, there may be a separate, soundproof *announce booth*, where an off-camera announcer can provide *voice-overs*.

The lighting control area houses the lighting console, which adjusts the settings of studio illumination, usually with multiple *presets* stored in memory for recalling different settings of light intensities, cross-fades, and so on. The *lighting director* uses the lighting console to remotely control each lamp in the studio to achieve the desired effect. The studio action is viewed on *picture* (or *video*) *monitors* in the control room, showing the output of each camera. The *vision* (or *video*) *engineer* also looks at the picture monitors and the *waveform monitors* for each camera, and adjusts remote camera controls to produce good-looking pictures that are also technically correct.

The audio control room or area has an audio mixing console and associated equipment, where the sound supervisor or *mix engineer* is responsible for producing the audio mix from the studio microphones and other sound sources.

Production control contains one or more desks with control panels for the *video switcher*, operated by the *technical director*, and other equipment such as a *character generator* and *graphics system*. This is also where the producer and other production staff sit. A bank of picture monitors, often mounted in a *monitor wall*, shows the output of each studio camera and other video sources, with two larger monitors for the Program and Preview outputs.

Most other equipment related to the studio is usually mounted in equipment racks in an *apparatus room* or *central technical area*.

Communication Systems

The various control rooms, and in some cases, individual stations or desks within them, are equipped with *intercom* terminals, to allow production staff to receive instructions from the program's producer or director, and to communicate with one another. Camera operators and other production staff (such as floor directors, who cue and otherwise manage on-camera talent) in the studio are also supplied with intercom on headphones.

When remote locations are involved in the production, those facilities or vehicles are also included on the intercom system. Typically the intercom provides two or more independent channels of communication, allowing all tech staff to hear the overall technical direction feed from the program's director, as well as various private lines ("PLs") that can be established for one-to-one communication between technical staff members at individual intercom stations.

On-camera talent is also equipped with another kind of communications system, usually supplied on small earphones worn in one ear only, and designed to be minimally visible on camera. Unlike the intercom system, the talent communication system typically includes the program audio mix (or a sub- or superset of it) on a full-time basis, with occasional cues added from the director. This is called *interruptible foldback*, commonly referred to as *IFB*. Remote on-camera talent (such as a field reporter on a news program) is also supplied with IFB, so communication lines must be established with remote sites to allow production staff and studio talent to communicate with remote talent via IFB.

These so-called "comm" systems (intercoms and IFBs) can be quite sophisticated in today's studios, especially for live productions such as news and sports programs, and although unseen and unheard by the audience, they are an essential part of any successful broadcast production.

SYSTEM CONSIDERATIONS

It will be apparent from the previous section that having to deal with pictures as well as sound produces a major increase in complexity for a television studio over a radio facility. This difference extends all the way through the television production system. As with radio facilities, some basic considerations apply to all television systems, including the following points.

Standard or High Definition

Most prime-time network programming is produced and distributed in high definition (HD) format. Increasingly, at the time of writing, all other forms of network and local programming in the U.S. is produced in HD, as well, with a dwindling amount of production in standard definition (SD). During the digital television transition in the U.S., local TV stations fed both analog NTSC and digital ATSC transmitters, but since 2009, only ATSC digital transmitters are used. Nevertheless, studio facilities may still accommodate SD program feeds (now typically in digital form, however) for one or more additional *multicast* program outputs for ATSC DTV. As discussed in Chapters 5 and 16, DTV video may be SD or HD, so even when stations are equipped to produce their local content in HD, they must still be able to handle SD content. Except where indicated, all video equipment described in the following sections is available in either SD or HD versions.

Analog or Digital

The trend in television broadcasting is to replace nearly all analog equipment with digital, and this trend is nearing completion throughout the industry at this writing. This has many advantages, particularly for recording and editing video signals, but also throughout the program chain, and is virtually obligatory for high definition systems. Some local TV stations may still have NTSC systems for feeding local cable systems or other specialized uses based on composite analog distribution. In some cases, certain studio equipment may remain analog (particularly for audio), but the outputs of these components are ultimately converted to digital to feed the ATSC DTV transmitter.

As with radio facilities, where such "analog islands" still exist within a digital facility, digital-to-analog and analog-to-digital converters are used on the analog unit's input and output respectively.

Compressed or Uncompressed Signal Distribution

As discussed under Bitstream Distribution and Splicing later in this chapter, there are alternative methods for distributing and recording digital signals within the broadcast plant, involving compressed bitstreams in place of uncompressed baseband signals.

File Transfer or Real-Time Video and Audio Distribution

As discussed in both the Video Servers and Ingest and Conversion sections later in this chapter, there is an increasing use of data files for moving program material about the facility, or among different facilities or servers in "the cloud" via IP interconnections. This arrangement uses computer networks instead of standard video and audio distribution, and it fundamentally changes the system design of a television broadcast facility. The term *file-based workflow* is often applied to such operations.

Stereo or Surround Sound

Stereo sound became the norm for NTSC programs in the latter years of analog television, and this has continued with the use of digital audio facilities for TV. ATSC DTV also makes provision for 5.1 channel surround sound, so most production and network centers now have capability for surround sound programming. Local stations are typically able to handle 5.1 audio channels in externally produced programs, although at this writing, only a minority of stations is equipped to produce local programs with surround sound.

Much of the audio equipment used in television and radio studios is generally quite similar, so we will concentrate here largely on the video aspects of television. The section on Audio for Television covers the main differences in equipment and systems incorporating special elements for television sound.

STUDIO SYSTEM

Figure 9.2 is a much-simplified block diagram of the video signal paths in a TV production studio system, showing a video switcher with various program sources, including *cameras, video recorders* and *video servers, graphics system, character generator*, and possibly other local and remote sources. It will be apparent that the general arrangement is similar to the radio studio system covered in Chapter 8, but here we are dealing with video sources and signals.

Where a source, such as a video server, provides an audio feed as well as video, then the audio is fed to a separate audio mixer, which is controlled separately from the video switcher, and which also takes audio-only sources such

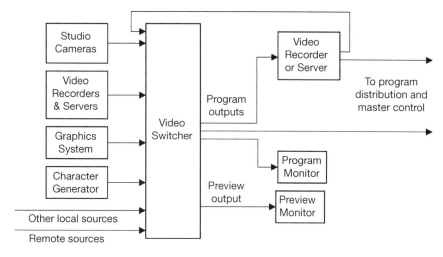

FIG. 9.2.   Television Production Studio Video Block Diagram

as announcer microphones, music, and sound effects to produce a complete program output.

As Figure 9.2 indicates, there are several outputs from the switcher, which are used to feed picture monitors, video recorders or servers, and program distribution systems. Not shown on the figure are the picture monitors that are provided for all program sources, or additional outputs for technical monitoring positions such as video, camera control, lighting, and sound operators.

POST-PRODUCTION EDIT SUITES

Traditional postproduction edit suites also use video switchers for putting together finished programs from prerecorded material, and may be arranged somewhat like a small studio control room. System arrangements are similar to those shown in Figure 9.2, except there are no live cameras and much of the equipment is controlled by an *edit controller*. See the Video Editing section later in this chapter for more information on video editing.

Let's look at the functions of the main components that make up a studio system.

Video Switchers and Effects Units

The heart of the system is the *video switcher*, also known as a *production switcher* and, in Europe, as a *vision mixer*. Figure 9.3 shows an example of the control panel for a fairly sophisticated switcher. The studio sources are fed to switcher inputs, from where they can be switched to feed multiple *buses*, using rows of push buttons as shown in the figure. This process is in fact a video version of the *routing switcher* shown in Figure 8.11 in Chapter 8. One switcher bus feeds direct to the *program* output, and the others feed one or more *mix-effects* (M/E) units. The M/E units enable various *transitions*, including *fade to black*, *cross-fades* or *dissolves*, and *wipes* (where the transition moves across the screen with different, selectable shapes). They also allow *video keying*, where parts of one picture may be inserted into another, and *chromakeying* (see following section). A *preview* output allows monitoring of any source or effect before it is actually put on air.

Switchers are defined by how many inputs and how many M/E units they have. Multiple M/Es allow more than one composite picture to be constructed and transitions to take place from one to the other. Many switchers have a *downstream keyer* (DSK) that allows a final key signal (often a character generator for captions) to be added to the switcher output, downstream of the M/E units. Switchers have basically the same operational features whether they are designed to handle analog or digital, SD or HD, video signals, although the signal processing inside is very different.

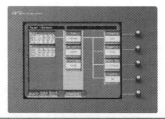

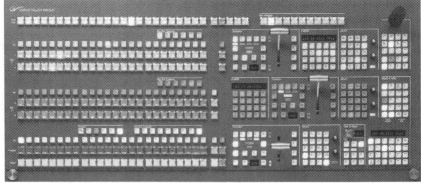

FIG. 9.3. Video Switcher Control Panel. Courtesy Thomson Grass Valley

Chromakey and Matte Systems

It is often desirable to combine two video signals so that one appears in front of the other. This occurs frequently in weather forecasts, when the presenter appears to be standing in front of, for example, a weather map or satellite picture. This is accomplished using a system called *chromakeying*, which is incorporated into most video switchers.

A chromakeying system takes one video signal and replaces every part of it that is a user-defined color (typically bright blue or green) with the corresponding video from a second video signal. The result is an output signal that combines selected portions of the two inputs. Let's look at how this works with a weather forecasting set.

On a typical set, the weather forecaster stands in front of a large blank wall or backdrop that is painted one solid color, say blue. The camera is focused on the forecaster, with the wall in the background, and the video from this camera provides the first input signal to the chromakeying unit. The second input signal is, say, the satellite video image from a weather graphics system. The chromakeying system takes every portion of the image from the camera that matches the blue color of the wall, and replaces it with the corresponding portion of the satellite image. Because the wall should be the only thing that is blue, this replacement creates the illusion in the output picture that the weather forecaster is standing in front of the satellite image. It is important that the forecaster's clothing, hair, skin, and so on should not match the color of the wall, or those elements would also be replaced, and the satellite image would show through.

A chromakeying system makes the weather forecaster's job a little tricky because he or she must appear to be pointing to portions of the background picture, while actually pointing to places on the blank wall. Typically, somewhere on the set, just out of view of the camera, there is a monitor showing the composite picture of the forecaster and the background image. The forecaster looks at this monitor to ensure that he or she is pointing correctly.

A slightly different system, which does not need a colored background, is known as a *matte system*. In this case, a separate image has to be produced to show which parts of one image should be replaced by a second image. It allows multiple colors in the first video signal to be replaced with video from the second video signal.

Digital Video Effects

Digital video effects (DVE) units enable zooming the picture size for picture-in-picture effects, squeezing it into different shapes, or changing perspective, as well as other effects. DVEs may stand alone or be built into the video switcher, and they may have channels for one or multiple pictures to be processed simultaneously. Some switchers allow sequences of effects with the M/Es and DVEs to be memorized and played back on demand.

DVEs work by storing every pixel of the TV picture in a frame store memory; using powerful computer processing, the pixels are manipulated to change their position in the frame to produce the effect required. Finally, a new frame is reconstructed and read out of the store as a regular video signal. This processing takes place extremely quickly, in *real time*, which means the pictures come out of the DVE as fast as they go in, with only one frame of delay, so the DVE can be used live on air.

Computer-based Post-production

While all of the post-production processes described above are performed by dedicated devices, it is increasingly possible to perform them within a single computer-based system, using specialized video production software. These systems can also incorporate non-linear video editing (see Video Editing section below), as well as animation and other advanced processes. In most cases, once the desired effects are applied within the computer system, it must be *rendered* to a standard video output format, a process that can take some multiple of real time to perform. In many cases, complete video productions from ingest of original content to output of finished work (including audio) can be done within a single computer-based system.

PICTURE AND WAVEFORM MONITORING

Picture Monitors

Picture (or video) monitors are used to check both program content and technical quality of the video signal. They are used extensively in production studios and throughout all broadcast facilities. The size of a monitor is usually expressed as the diagonal screen measurement, and sizes range from as little as 2 or 3 inches to about 36 inches for a monitor showing a single picture. Monitors may be either 4:3 or 16:9 aspect ratio, depending on the sources being monitored (e.g., SD is typically 4:3, and HD is always 16:9), and these days color monitors are nearly universal. In control rooms, they may be built into a monitor wall, but they are also frequently mounted in control desks or equipment racks in other areas.

Traditionally, cathode ray tube (CRT) displays were used for picture monitors, but more recently other display technologies have become the norm, including flat panel plasma and liquid crystal displays (LCD). Plasma and LCDs have advantages of lighter weight and smaller size than CRTs, but in their earlier forms they did not produce as high a picture quality, so CRTs were still preferred for critical technical monitoring applications in some cases. As the flat panel technologies have matured, however, and specialized, professional flat-panel monitors have developed, they have subsequently become the preferred form factor for nearly all picture monitoring applications.

Very large flat-screen monitors, many feet in width, either using plasma or LCD displays, are increasingly being used in production and master control rooms. In this case, rather than have many individual monitors for different sources, the large screen is divided up into multiple picture areas, giving the effect of a "virtual monitor wall." One advantage of this arrangement is that the picture sizes and allocations can be easily changed to accommodate different setups, and other information can also be included on the screen, such as source identification, audio level metering, clock, and alarm indicators for fault conditions.

Waveform Monitors

As the name implies, *waveform monitors* are used to measure the technical characteristics of the video waveform (see Chapters 4 and 5). In particular, they are used to check the black level and output levels from television cameras, because these parameters have operational controls that are continuously changed during television production. Waveform monitors are also used at different points in the program distribution chain to verify compliance with technical standards.

An analog waveform monitor shows the shape of the actual video signal as it varies in time. A digital waveform monitor re-creates that waveform to its familiar appearance from the stream of digital video data. Digital waveform monitors are able to perform numerous other checks on the digital video signal, as well.

Traditionally, waveform monitors were stand-alone items of equipment with a dedicated screen. Modern units often have a measurement section at the location where the video signal needs to be monitored and a separate monitor display, which may be at a different location, with communications between the locations performed over a computer network. Depending on the system, the display may also show other information about the television signal, including the picture itself. Some systems also allow remote adjustment of the video signal, for correction of black level, peak level, and other parameters.

Vectorscope

A *vectorscope* is a special sort of waveform monitor that shows only the video chrominance information in a *polar display* (parameters are shown as vectors, with a distance and angle from the center of the screen). Its primary use is with analog composite video, where it is important to check that the color subcarrier is correctly adjusted. Signals for each color appear at a different location on the screen, so the six primary and secondary colors of standard color bars have boxes marked on the screen for where they should be.

TELEVISION CAMERAS

Studio Cameras

Studio cameras are designed to produce high-quality video images, and they have many features for convenience and ease of use, including large *viewfinders*, the ability to have high-quality, large lenses with wide zoom ranges, and comprehensive monitoring and communications. When necessary, they can be fitted with *teleprompters*, which are special electronic displays of script for presenters to read while looking into the camera lens. It is not easy to provide all of these features on small, lightweight cameras, so studio cameras tend to be fairly substantial, as shown in Figure 9.4.

Studio cameras are connected through a *camera cable* to a *camera control unit* (CCU) in the equipment area associated with the studio control room. This unit provides power and communication links to the camera (so the operator can hear and talk to the staff in the control room using a headset). It enables the

FIG. 9.4.   Television Studio Camera. Courtesy Sony

camera output to be fed to the video switcher and allows for remote controls in the vision control room to adjust and match the appearance of the camera picture.

Portable Cameras

Small, portable cameras at one time sacrificed performance for portability and were used primarily for news gathering, where light weight and low power consumption were more important than picture quality. However, current generation portable cameras are now available for both SD and HD with similar performance to the best studio cameras. They are, therefore, being used in all aspects of television production inside and outside the studio. Most portable cameras have built-in (or can be fitted with) a video recorder of some sort (i.e., they are a camcorder) and are powered from a battery, so they are portable and self-contained. This arrangement is universal for news gathering and most field production (see Chapter 10).

When used in a studio, portable cameras are fitted with an adaptor, allowing use with a camera cable and CCU, as with a studio camera. They may be handheld or used with camera support devices, and they may also be fitted with special mounts for large lenses and large viewfinders, thus providing many of the capabilities of studio cameras.

Lenses

Most imaging starts with light passing through a lens. Television cameras use *zoom lenses* that have adjustable focal lengths. At the *telephoto* end of the zoom range, the camera sees a small amount of the scene, which is enlarged to fill the frame. At the *wide-angle* end of the range, it sees much more of the scene, as though from farther away.

The camera operator adjusts the lens zoom and focus settings with controls mounted on the *pan bars* at the back of the camera, while looking at the camera picture in the viewfinder. There may be a *shot-box* that allows the operator to preset different lens settings and go instantly to them at the push of a button. The *aperture* of the lens is adjusted by the video engineer in the control room.

Camera Imaging

The scanning process for video images was explained in Chapter 4. To recap briefly, when a TV camera is pointed at a scene, the image passes through the lens and is focused onto a light-sensitive pickup device of some sort. The target or sensor produces variations in the electric charge-density level; it is scanned to dissect the picture into scan lines, and this process is repeated to produce sequential fields and/or frames. Based on the amount of light detected at any point in time, the camera produces a voltage that is proportional to the amount of light at each point in the picture.

Note that each frame in the camera is scanned bottom to top, whereas in a picture display device it is top to bottom. This is because the image produced by the camera lens is actually upside down, so the scan also has to be inverted.

The device that did the image scanning in older television cameras was a *camera tube*, or *pickup tube*. An electron beam actually scanned the electric charge on the inside surface of the tube to create the electronic version of the video image. In most modern cameras, the image pickup device is called a *charge coupled device* (CCD), or a similar solid-state sensor (more below).

CCDs for cameras use a pair of electronic matrices (a two-dimensional array), as shown in principle in Figure 9.5. Each matrix comprises a very large number of individual photosensitive "cells." The number of cells determines the resolution of the camera, and many more cells are needed for an HD camera compared to SD. Each cell in the first matrix is charged up to a level that is proportional to the amount of light falling on it. The individual charges from each of the cells in the first matrix are transferred simultaneously 30 or more times per second to corresponding cells in a second matrix, which is shielded from the light. The circuitry in the camera then scans the second matrix, and the samples from each cell are processed to produce the video signal.

The charges in the first matrix must be transferred to a second matrix before they are scanned because the first matrix charge levels are constantly changing as the image through the lens changes. So, a "snapshot" of the first matrix must be created to ensure that the correct fixed image is converted to an electrical signal by the camera. Depending on the camera, the CCDs may be scanned to produce an output image with 4:3 or 16:9 aspect ratio, or in some designs it is switchable to either format.

The best-performing CCD color cameras, as used for broadcasting, actually use three CCDs—one each to scan the electric charge image created by the red, green, and blue light coming through the lens. As shown in Figure 9.6, the light from the lens passes through an optical *beam splitter*, based on glass

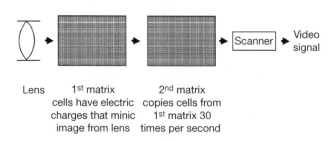

Lens   1st matrix   2nd matrix
            cells have electric   copies cells from
            charges that minic   1st matrix 30
            image from lens   times per second

FIG. 9.5. CCD Imaging Process

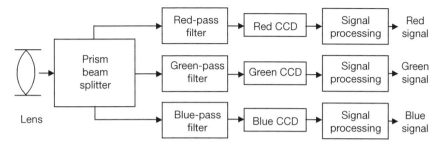

FIG. 9.6. Basic Design of a Three-CCD Camera

prisms, which splits the incoming optical image into three different color beams (red, green, and blue). The beams then pass through additional filters, which remove any remaining unwanted colors, and the purified individual color images fall onto their own individual CCDs, where they are scanned, as described previously.

These are typically called "three-chip" cameras, a broader term that incorporates those units that use a solid-state image sensor other than the CCD. The most common alternative pickup device is based on Complementary Metal Oxide Semiconductor (CMOS) technology. This enables cameras that are smaller in size and lower in power consumption. It is likely that both CCD and CMOS devices will be used in television cameras for some time to come.

The three separate color signals that come from the three pickup devices are *component video* signals. These are further processed and may be fed out of the camera as three separate analog component signals. However, most cameras carry out analog-to-digital conversion and produce a serial digital video output (SD or HD, depending on the camera type), as described in Chapter 5. An SD camera may combine the analog component signals to produce an NTSC composite video signal, as described in Chapter 4.

HD cameras often provide a *downconverted* digital SD or composite analog video output as well as the HD feed. This provides extra flexibility in how the camera signals may be used.

Camera Support

Portable cameras may be handheld (most actually rest on the cameraperson's shoulder), but, whenever possible, cameras are fixed to some sort of mount. This is less tiring for the operator and allows for pictures without *camera shake*. Both portable and studio camera types may be mounted on a *pan-and-tilt head*, which allows the camera to be smoothly pointed in the right direction. In studios, this head is mounted on a wheeled *pedestal*, which adjusts smoothly up and down, or on a tripod with a wheeled *skid*, both of which allow the camera to be *tracked* across the floor. For special high- or low-angle shots, cameras may be mounted on an elevating *crane* or a counterbalanced *boom*.

When a portable camera is required, but smooth movement akin to studio camera operation is desired, a specialized mounting system can be used that allows the handheld camera to be stabilized, through the use of gyroscopic mounts and other techniques. This approach to camera mounting is often referred to by the name of an early and popular brand of such stabilization mounts called *Steadicam*™, although other brands are available. Modern cameras also may provide the option of electronic image stabilization via special processing of the video signal inside the camera prior to output.

Some news studios do not use a camera operator, so the camera is mounted on a *robotic pedestal* with remote controls in the control room for camera positioning and lens zoom and focus.

Isocam

In a live production environment, the various cameras are selected and switched through the video switcher, to produce the final output. This output can be recorded for subsequent editing and replay. An alternative arrangement is often used where multicamera shows are recorded for subsequent editing. In that case, the output of all cameras is recorded simultaneously, so they are all available during the postproduction process. This arrangement is known as *isocam* (presumably from "isolated camera").

FILM IN TELEVISION

Acquisition

The aforementioned process describes the use of electronic cameras to acquire video images. However, some television programs are actually shot with film cameras, both in studios and outside on location, using techniques similar to those used for making cinema films. Some directors prefer working with film, and some television executives just prefer the "look" of film-produced pictures. This trend is diminishing as modern digital video cameras allow this look to be accomplished electronically, and with the greater convenience and reduced cost of video. (Even some cinematic productions have moved to video for initial image capture.) Nevertheless, film origination still exists, so for completeness and historical perspective, it is described below.

Most productions shot on film use 35 millimeter film cameras for the highest quality. Low-budget productions, and especially those shot in remote locations with portable cameras, may use 16 millimeter film, with smaller and lighter cameras and lower film costs. Originally, film was edited by cutting and splicing the negative, with special effects produced in optical printers. The finished film was then played out or transferred to a video recording using a *telecine* machine.

Subsequently, however, most film material was transferred to video soon after processing of the original reel. From postproduction onward, the process was then the same as for video-produced material, meaning that editing was done electronically using video editing systems, rather than physical splicing.

Thirty-five millimeter film can produce excellent pictures for both SD and HD video, while 16 millimeter film is usually more suited to SD, because its resolution limitations, marginal image stability, and film grain can result in image quality that is inferior to electronic HD pictures.

Telecines

A *telecine* scans film frames and produces a regular video signal output. One type of telecine, used for many years, was basically a film projector mated to a device like a television camera. Instead of the optical image from the projector falling on a screen, it was directed into the camera lens and each frame was scanned while the film was stationary in the *film gate*. Some types of modern telecines split the light from the film into red, green, and blue beams and use three *area-array CCD pickup devices* to scan each frame. Other telecines use three *line-array CCD* devices to detect the light shone through the film frame, after it has been split into red, green, and blue components. With this system, the film moves continuously past the scanner, and the motion of the film produces the effect of vertical scanning as described for NTSC video in Chapter 4, while the horizontal scan is performed by the row of CCD cells. An older system, which is still used in some facilities, is the *flying spot scanner*. In this design, the light shining through the film comes from a moving spot of light emitted by a high-intensity cathode ray tube (CRT). This "flying" spot scans the film as it moves continuously past the scanner. The motion of the spot and the motion of the film together produce the effect of horizontal and vertical scanning. The light is split into red, green, and blue beams and detected by three *photomultiplier* devices. In both CCD and flying spot telecines, considerable signal processing is required to produce the final video output signal, usually including an electronic frame store. Most telecine transfers also include the process of *color correction*, to correct color casts and other anomalies of the pictures on film.

At one time, telecines were widely used by networks and broadcast stations for broadcasting feature films and news shot on film. Most stations no longer use such machines, but these are still used by post-houses and archives for transferring film to video.

Film Scanners

The best quality high-definition film transfers are done with a *film scanner*, rather than a telecine. This device uses CCDs with very high resolution and produces a digital data file output rather than a video signal. The transfers may take place in slower than real time, thus allowing very high quality to be achieved. The output file is then available for postproduction work.

3:2 Pulldown

Film runs at 24 frames per second, while U.S. television video uses approximately 30 frames per second. A system called *3:2 pulldown* is used so that film images can be transferred to video. You will remember from earlier chapters that NTSC and interlaced digital video use two fields for each frame so there are 60 fields per second. In the 3:2 system, the first film frame of any consecutive pair is converted by the telecine into three video fields, while the second film frame is converted into two video fields, and so on consecutively. The two film frames occupy 2/24 of a second, which is 1/12 of a second, and the five (3 + 2) video fields occupy 5/60 of a second, which is also 1/12 of a second; so the timing works out exactly over any five-field period. A similar process is applied to 60 fps non-interlaced digital video transfers, in which case the two film frames are converted to five complete video frames (rather than interlaced fields) in such a 3:2 sequence.

One result of the 3:2 pulldown sequence is that a moving object in the picture may exhibit a slight but noticeable judder, because alternate film frames are each effectively displayed for different lengths of time. Some modern DTV receivers can perform a process called *reverse 3:2 pulldown* that effectively reverses the process and allows for smoother motion to be portrayed, with a progressive scan image.

VIDEO RECORDING

Videotape Recorders

Videotape recorders (VTRs) are rapidly being replaced by hard disk-based servers in many studio applications and by optical disk and solid-state flash memory recorders for portable use. There was, however, a large installed base of different types of VTRs throughout the industry over many years, and this form of storage formed the backbone of television operations for a long while. Many are still in use. For this reason, and for historical perspective, the topic is covered in some detail below.

VTRs have some similarities to their audio counterparts, and they all use magnetic tape as the media on which the program material is stored. Early VTRs used tape stored on open reels, but all modern machines house the tape in a cassette for protection.

As well as recording video signals, VTRs record two or more channels of audio information and a *timecode* signal (see later section on SMPTE timecode). A control track is also usually recorded on tape for machine synchronization during playback. Arrangements vary somewhat, depending on the particular format and, in particular, whether the VTR uses analog or digital recording.

Different types of VTRs are defined by the video format(s) they can handle and the *tape format* used. This includes not only the physical characteristics of the tape but also all aspects of the way the signals are recorded. If a recording is made on one machine with a particular format, it should play back on any other machine of the same format. Many different tape formats were developed at different times and by different manufacturers, each with their own characteristics as summarized in the sections for analog and digital VTRs below.

Magnetic Recording

As with audio, video signals are recorded onto tape with a *record head* that converts the electrical signal voltage into a varying magnetic field. This magnetizes the tape coating as the tape moves past the head. Upon playback, the head picks up the stored magnetic information and converts it back into an electrical signal voltage.

The higher the frequencies that have to be recorded, the higher the speed at which the tape must pass the record and playback head. Because video signals have much higher frequencies than audio, a much higher tape speed is needed. Digital video signals contain higher frequencies still, and need even higher speeds. However, it is not practical to pull the tape past a stationary head at a high enough speed, so the problem is overcome using a system called *helical scan*, also known as *slant track* recording.

Helical Scan

With the exception of the obsolete quadraplex VTR, all practical VTRs rely on *helical scan* to achieve the high head-to-tape speed needed for video recording. Details vary, but the principle is the same for all machines, as shown in Figure 9.7.

Two or more record and playback heads are fixed to the outside of a rapidly rotating drum, around which the tape is wrapped at an angle, like part of a spiral helix (hence, the name). The tape moves quite slowly past the drum, but each head moves rapidly across the surface of the tape, laying down a series of diagonal tracks. (The angles shown in the figure are exaggerated for clarity, so the tracks on the tape are actually much longer than shown.)

Because it is not practical to record video signals directly onto tape, various modulated carriers are used for different VTR formats, digital and analog

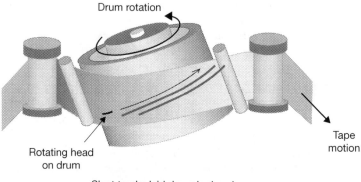

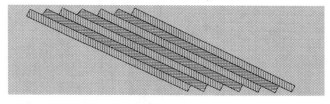

FIG. 9.7. Principle of Helical Scan Recording

being quite different. The video information on the slant tracks, recorded by the *rotating heads*, takes up most of the tape area. There may also be fixed heads that record and playback audio, control, and timecode signals on narrow *longitudinal tracks* along each edge of the tape (not shown in Figure 9.7). These tracks vary with different machines and, in some cases, the audio and/or timecode signals may also be recorded on the slant tracks using the rotating heads. In the case of digital VTRs, the audio is typically embedded with the video bitstream and recorded to tape with the rotating heads.

Timebase Correctors

Videotape recording for broadcasting would not have become possible without the invention of the *timebase corrector* (TBC). Because a VTR is a mechanical device, there are slight variations in the speed of the tape and the rotation of the drum. These produce irregularities in the timing of the replayed video signal, and the sync pulses and picture information may arrive at slightly the wrong time. The timing variations are not important if the signal has only to feed a television set direct (as a consumer VCR does). In that case, the set will remain locked to the sync pulses even as they vary and will produce a basically stable picture. Such timebase errors are disastrous in a broadcast station, however, where the signal has to be mixed with other signals, perhaps edited, and re-recorded. To make VTR output signals usable, they are passed through a TBC that comprises an electronic buffer store. This removes the input irregularities and sends out video with stable timing restored.

Digital TBCs offer a much greater correction range than analog and are universally used, even with VTRs that otherwise record an analog signal on tape. Digital VTRs do not actually have a separate TBC function, however, because the techniques required to recover the digital data from the tape are inherently time-stable, and thereby take care of correcting any timing irregularities.

In approximate historical order of introduction, we list below the principal videotape formats you may come across, with their main characteristics. Some of the machines mentioned are now obsolete and may be found only in specialist facilities for dubbing archive material onto more modern formats, but they are included for completeness. (There are a few lesser known formats that are not listed.) With the exception of the open-reel Quadraplex and Type C formats, all VTRs described below use tape enclosed in a cassette for protection (with two reels inside the cassette).

Slow and Fast Motion

Most VTRs are able to produce pictures at speeds both slower and faster than when they were recorded, by speeding up and slowing down the tape. There is a limited range of speeds over which the pictures retain broadcast quality, depending on the format.

Analog VTRs

Two-Inch Quadruplex (1956)

Developed by Ampex Corporation in the United States, and also known as *Quad*, this format is now completely obsolete and is mentioned here only because it was the first successful VTR format. The name comes from the wheel with four heads, which rotated across the width of the tape to produce the video tracks. The system used open reels of two-inch-wide tape, which were extremely heavy and cumbersome, but at one time were the means by which nearly all television program material was stored or shipped from one place to another.

U-Matic (1970)

The U-Matic format, developed by a Japanese consortium, used 3/4 inch tape in a cassette shell. Video quality was not particularly high, but it improved considerably with the later High Band version. The format may still be found in use by some stations and production houses. Portable versions were once widely used for news gathering, as a replacement for 16 millimeter film.

Type C (1976)

The Type C format, developed by Ampex in the United States and Sony in Japan, used one-inch tape on open reels to record composite video, with two channels of audio. It was the first machine to provide usable still frame, slow- and fast-motion playback, and pictures in shuttle. At one time, this was the most widely used VTR format, but it has now disappeared from regular use.

VHS (1976) and S-VHS (1987)

Developed by JVC, VHS (Video Home System) is the very successful standard format for consumer VCRs, using a 1/2 inch tape. It was widely used in homes and businesses throughout the world, and although largely replaced by the DVD format, it may still be found in some facilities.

The S-VHS version had higher resolution, delivered better color performance, and could record timecode. It was also very cost effective, so it was used by some small-market broadcasters, but was later superseded by other, higher-quality professional formats.

Betacam (1982) and BetacamSP (1986)

The Betacam format, from Sony, uses 1/2 inch videotape to record analog video. Input and output interfaces use composite video, but the signal is actually recorded on tape in analog component form. This considerably improves picture quality. BetacamSP (superior performance), with different tape and higher recording frequencies, further improves the quality. BetacamSP became the industry workhorse for portable recording and studio applications before digital formats were introduced, and some may still be in use today.

Video8 (1983) and Hi8 (1989)

These formats, using 8 millimeter tape, were developed by Sony for consumer camcorders, in conjunction with several other manufacturers. Hi8 was occasionally used for professional assignments requiring a very small camcorder, before the introduction of digital MiniDV.

M (1983) and M-II (1985)

These formats from Panasonic, using 1/2 inch tape, were developed for professional use from the VHS format. They were direct competitors to Betacam and BetacamSP but were not widely adopted and are now obsolete.

Digital VTRs

Where video tape recording is still used, digital VTRs have largely replaced analog machines due to their superior video and audio performance and other advantages. One crucial advantage of digital VTRs is the ability to record many *generations* of video without degradation. This is important with editing and postproduction, where the original material needs to be re-recorded many times. Analog recordings show a buildup of video noise and other degradations after a very few generations, which does not occur with digital. Figure 9.8 shows an example of a standard definition Digital Betacam VTR.

Video Compression for Recording

It is possible to record video signals onto tape in uncompressed form. Uncompressed video has a high data rate (270 or 360Mbps for SD, 1.485Gbps for HD), however, so many digital VTRs reduce the amount of data by using *video compression*, while trying to maintain quality as high as possible. This allows smaller tape sizes and smaller head drums to be used, which can make the machine more compact, and also allows less tape to be consumed per unit of time. MPEG-2 video compression, as used for ATSC transmission (see Chapter 16), is used for some VTRs, but with lower compression ratios than used for transmission. Several other compression systems are also used for recording.

The digital VTRs listed first are all designed to record standard definition video only.

FIG. 9.8. Studio Digital VTR. Courtesy Sony

D1 (1987)

The first digital VTR was developed by Sony to record the new 601 component video format defined by the CCIR (later known as ITU-R BT.601—see Chapter 5 on digital video). It used 19 millimeter tape, with a recording time up to 76 minutes of uncompressed video. The machine was expensive and used mainly for post-production, where its multiple generation capability was important. It is no longer widely used.

D2 (1989)

Produced by both Sony and Ampex, the D2 format was unusual in that it recorded composite NTSC (or PAL) video in digital form. It used 19 millimeter tape, with up to three hours of record time. While allowing easy integration into existing analog facilities, it was eventually overtaken by component video recorders.

D3 (1991)

The D3 format came from Panasonic and also recorded composite video in digital form. It used 1/2 inch tape, with up to four hours of record time. With an alternative small cassette, it was possible to make equipment suitable for electronic news gathering (ENG) and other portable applications.

Digital Betacam (1993)

Sometimes referred to as "DigiBeta," the format was developed by Sony, with some decks also able to play back analog BetacamSP tapes. It used light 2:1 compression to record component digital video. Quality was high, and the machine became the industry standard for high-quality standard definition recording for portable and studio applications for a substantial period. It is still in use in some facilities.

D5 (1994)

D5 was produced by Panasonic as a successor to D3, and it recorded component video with four audio channels onto 1/2 inch tape. It used a higher data rate than Digital Betacam and recorded uncompressed video at up to 360Mbps for the very highest quality, but had the disadvantage of not being available in a portable recorder version.

DV and MiniDV (1995)

Developed by a consortium of ten companies, this format was intended for consumer camcorders. It records video and two audio channels onto 1/4 inch tape, with 5:1 compression using the DV codec. Maximum recording time in standard mode is 80 minutes. This format became widely used for professional products.

DVCPRO (1995) and DVCPRO50 (1998)

DVCPRO from Panasonic was a development of the DV format for broadcast applications, using a wider track and a different tape material to increase robustness and quality. It also records longitudinal cue and control tracks for improved editing performance. This format was widely used by broadcasters, particularly for ENG, and had a good range of portable and studio recorders and editing facilities. The later DVCPRO50 further improved quality by recording at a higher bit rate (50Mbps), at the expense of reduced recording time.

D9 Digital-S (1995)

This format, developed by JVC, used a 1/2 inch tape cassette, similar to VHS, but is, in fact, a much higher-quality professional machine, rivaling Digital Betacam. It uses a variant of the DV codec, with some similarities to that used for DVCPRO50, but with a different tape format, and was available in both portable and studio versions.

DVCAM (1996)

This is the Sony professional version of DV, with quality and capabilities similar to DVCPRO.

Betacam SX (1996)

This format from Sony was targeted for ENG and newsroom applications, using quite heavy 10:1 video compression. Betacam SX VTRs were the first to allow video to be sent back to the studio over video links at twice the normal speed ("2x real time").

Digital8 (1999)

This format from Sony recorded the same digital signal as DV onto less expensive Hi8 tapes, and could play back analog Video8 and Hi8 tapes. Intended as a consumer camcorder format, there was also an ultra-compact editing deck and the format was occasionally used for very low cost professional applications.

IMX (2000)

This Sony format had some similarities to Digital Betacam but used MPEG compression and could record eight audio channels. As well as regular playback of video and audio, the recorded compressed data could be output directly from the tape, for transferring to video servers.

HD Digital VTRs

These VTRs are all intended for recording high definition video signals with either 1080 or 720 lines of resolution.

D5 HD (1994)

This HD format from Panasonic was based on the SD D5 recorder and used 4:1 Motion JPEG video compression to record video and four audio channels (eight channels on later models). It provided high-quality HD recordings and was widely used as a mastering machine, but was available only as a studio recorder.

HDCAM (1997)

This Sony HD format was based on the Digital Betacam recorder mechanics. Compared to D5, this format had more aggressive compression, of about 7:1 and, in order to reduce the data rate sufficiently for the tape format, it subsampled the video to a horizontal resolution of 1440 pixels for the 1080-line format, compared to 1920 for D5 HD. It was available in both studio and portable versions and was widely used for high-definition program acquisition.

DVCPRO HD (2000)

This HD version of the DVCPRO format from Panasonic recorded video with considerably higher compression and more aggressive subsampling than used for either D5 or HDCAM, to achieve a bit rate of 100Mbps recorded on tape. It also subsampled the video to a horizontal resolution of only 1280 pixels for the 1080-line format.

D9 HD (2000)

This HD version of Digital-S from JVC recorded video and up to 8 channels of audio at a data rate of 100Mbps, using a codec developed by JVC.

HDCAM SR (2004)

This later version of HDCAM, from Sony, recorded HD video at a data rate of 440Mbps with full 1920 × 1080 resolution, using mild compression of only 2.5:1, and with up to two hours of recording time. It carried 12 channels of uncompressed audio and was intended for the highest-quality production and distribution applications.

HDV (2004)

Sony, JVC, Canon, and Sharp jointly produced the HDV standard for consumer and professional use. HDV recorded an HD signal to MiniDV tapes using heavy MPEG-2 compression and aggressive subsampling to permit recording at 25Mbps, with a recording time of 60 minutes. The format is based on technology previously developed by JVC. Both camcorders and studio recorders were produced.

Optical, Solid-State, and Hard Disk Video Recorders

Optical Disk Recording

In 2003, Sony introduced the XDCAM recording format, which used an optical disk for recording SD or HD digital video and audio. The disc was similar to, but not the same as, Blu-ray DVD and could be recorded, erased, and re-recorded. SD signals were recorded with a choice of compression formats, including DVCAM and IMX, with up to 85 minutes of program. Files could be accessed and transferred direct from the disc to other systems such as video servers. XDCAM was available as a camcorder and studio recorder and was intended to be part of an integrated system for managing digital video content.

Subsequently other systems using optical disk as a portable video recording medium emerged, in a variety of form factors.

Hard Disk

Numerous manufacturers have produced portable camera systems that record digital video and audio direct to a hard disk. The recording capability can be either built-in to the camcorder, added to a camera as a docking module, or connected to a camera as a standalone recorder. The recorded files can be transferred at high speed from the hard disk direct to a video server or nonlinear editing system (see below).

Solid-State Memory

Panasonic introduced the P2 recording format in 2003, which used solid-state memory for recording either SD or HD digital video and audio, and had the advantage of containing no moving parts. The recorder used up to five Secure Digital flash memory plug-in cards. Signals were recorded with DV or DVCPRO compression. P2 was available as a camcorder and studio recorder. Both supported MXF file transfer (see later section) and were intended to be part of an integrated system for managing digital content.

Many other products using solid-state memory have followed, with the once again available recording capability either built into the camera, dockable to a camera, or as standalone portable recorder. At this writing such systems are becoming quite popular for portable video recording. Their use of standard, widely available, removable recording media and their fast access to stored content are key advantages. (Note that a popular storage medium here is the Secure Digital card referred to above, and its family of more recent extensions, all of which use the "SD" abbreviation, which should not be confused with the use of the same abbreviation for standard-definition video.)

VIDEO EDITING

In the early days of quadraplex VTRs, videotape was edited by actually cutting and splicing the tape. This crude system soon gave way to editing by copying (*dubbing*) portions of the source VTR material onto a second VTR. This method provides *cuts-only* edits, without any other types of transitions. Later VTR editing suites typically used several playback VTRs for the program material sources (known as *A-roll*, *B-roll*, and so on), fed through a video switcher to allow transitions with mixes, wipes, and effects to be carried out. Graphics, captions, and other elements could also be added at the same time. The output of the switcher with the finished program is fed to a recording VTR.

All of the machines and the switcher are controlled by an *edit controller*. This system locates the right program material and tells the VTRs when to start and stop and when to record. The controller monitors the tape *timecode* to know exactly which piece of program material is playing at any particular time.

The edit controller uses an *edit decision list* (EDL), which has all of the instructions for assembling the program. This includes which tape reels are needed and the particular timecode readings for the beginning and end of each segment to be included in the program. EDLs may be prepared in the edit suite, but they are frequently prepared *off-line*, using copies of the program material recorded on a low-cost tape format such as VHS or MiniDV, with *burnt-in timecode*.

Such facilities have largely been replaced by nonlinear editing (NLE) systems, where many of the effects previously done with a video switcher can be achieved more efficiently using the NLE system software. It is also possible, however, to use some edit control systems with both VTRs and *video servers*, enabling integrated postproduction with both types of recording system.

SMPTE TIMECODE

For video production, editing, and transmission, it is necessary to time program segments accurately and also to be able to identify any single frame of video. For this purpose, the Society of Motion Picture and Television Engineers

(SMPTE) developed a system called *SMPTE Time and Control Code*, usually known as *timecode*. This digital code is recorded onto videotape to identify how many hours, minutes, seconds, and video frames have passed since the beginning of the recording. On early analog VTRs, timecode was recorded on a separate track along the edge of the tape—this is known as *linear* or *longitudinal timecode* (LTC). The disadvantages of LTC are that it takes up extra space on the tape and cannot be read when the tape is stationary or moving slowly. A later development put timecode into the vertical interval of the video signal (see Chapter 4), where it is known as *vertical interval timecode* (VITC, often pronounced "vitsee"). VITC can be read at most tape speeds (whenever the video is viewable), which is a great advantage during editing or when cueing a tape for broadcast. For maximum flexibility, some VTRs and editing systems can use both LTC and VITC.

SMPTE timecode is recorded when the video is first recorded; it can then be used to identify edit points during postproduction. If necessary, a new continuous timecode is laid down for a finished program. This can then be used for timing and machine control purposes, in conjunction with an automation system, when the video is played out for broadcast.

Timecode is also used with video servers and nonlinear editing systems. It has many other applications: it can, for example, be used to synchronize a separately recorded audio recording with a video recording, control station clocks, and act as the time reference for station automation.

Burnt-in timecode is used where an editor, producer, or other viewer needs to view program material and also see the timecode at the same time, to determine edit points. The timecode numbers are keyed into the video, so they will always be visible, allowing a copy to be made onto a low-cost tape format such as VHS or MiniDV, which can be viewed *off-line*, away from the editing suite facility.

VIDEO SERVERS

Chapter 8 described how hard disk recorders for audio have revolutionized radio broadcasting. The equivalent development in television is the *video server*, which is widely used for recording, editing, storing, and distributing digital video signals. A video server is basically a powerful computer that is capable of receiving digital video signals, storing them on hard drives, and playing them back when required. As with audio servers, video servers can playback material while still recording, and stored material can be played back at random by multiple users simultaneously.

Recording Formats and Interfaces

Although called video servers, these devices in fact record both audio and video. They accept signals over standard video and audio interfaces but, like all computer data, the video and audio is stored on the hard disk system in the form of files. The video may be stored uncompressed (sometimes known as *linear* recording) or, to reduce the size of the files, it may be stored in compressed form (particularly for HD format signals, which produce very large files). Several systems are used for video compression, largely based on the same methods used with digital videotape recorders.

When a digital video file is retrieved from a server system and sent, say, to a video switcher, the server reads the file from the disk, formats the video and audio into some form of standardized serial data stream, and sends the stream(s) out to the switcher—either as separate audio and video signals, or as a single signal with audio *embedded* with the video. Servers also have a network interface for receiving and distributing files over a LAN or WAN.

Server Types

There are many varieties of video servers. Some are based on standard computers, with the usual keyboard, mouse, and monitor for the operator interface. However, most broadcast servers use custom hardware optimized for video applications, and with provision for all of the video and audio input and output interfaces. Some versions provide front-panel controls with buttons, knobs, and indicators for specific functions, similar to a videotape recorder.

Video servers may be associated with a production studio, a postproduction facility, an on-air news studio, or a master control switcher for transmission. In most cases, they are connected to a computer network and hence, able to share files. A large video server system, therefore, can provide the functions of both a library and a distribution system. An example of a server-based playout system is shown in Figure 9.9 later in this chapter.

Archiving

Although server systems can record large amounts of programming on hard drives, when a station puts its entire program library onto the system, it may run out of hard drive capacity. To efficiently store programming for archive purposes, the material may be transferred to special magnetic data tapes, which can store very large amounts of data for long periods with great reliability. The disadvantage of this storage method is slow access to any particular program item.

File Interchange and MXF

The file formats used by different servers are often different, depending on the manufacturers of the particular system. This can make interchange of program files among producers, networks, and stations rather difficult. A standard known as MXF (for *Material eXchange Format*), makes such interchange easier. MXF enables systems from different vendors to "talk" to each other for exchange of video and audio material, and to include all the *metadata* (information about the program material) that is needed to make optimum use of the program material.

Slow and Fast Motion

Servers are well suited to producing video with either slow or fast motion because recorded frames can be easily processed without concern for retrieval from tape at nonstandard speeds. Systems are available that are specially designed for playing back slow and fast motion. These playback speeds, as well as visual scrolling in rewind and fast-forward of file-based video, are sometimes referred to as *"trick modes."* The quality of slow-motion images is also improved considerably by using a camera that produces images with higher than normal frame rates. Video intended for slow-motion playback—such as sports replays—is therefore typically recorded at several hundred frames per second.

NONLINEAR EDITING

As with digital audio, which can be edited on a PC or dedicated audio workstation, video nonlinear editing may be carried out on a high-power general-purpose PC fitted with an editing software program, suitable input and output interface cards, and sufficient hard disk storage. Alternatively, some sophisticated editing systems use dedicated custom hardware, although at their core, they all incorporate computer processors and hard disk storage. In all systems, program material can be selected, cut, reordered, modified, and generally prepared for transmission.

Advantages of disk-based systems, compared to traditional tape-based editing, include the ability to instantly access any program segment, and to go back and revise edits made earlier in the program, without affecting later edits. Such *random access* features allow *nonlinear editing*, which is in sharp contrast to having to shuttle a tape backward and forward to locate a particular program segment.

All nonlinear editors provide at least some of the functions of both a video switcher and audio mixing console, but implemented in software. This allows pictures and sound to be modified, and several sources to be mixed and combined together, often with special effects, all carried out on the computer. In fact, many functions for television postproduction, which previously needed a room full of equipment, can be done on a PC with a few peripheral items of hardware. This has allowed new ways of programming and has blurred the definition of what is really needed for TV production and postproduction.

Digitizing

Depending on how the program material is originally recorded, it may be necessary to transfer it to the editing system in a process known as *digitizing*. The video is played back and passed through a processor, which records it to the system's hard drive in a suitable compressed or uncompressed digital format. Many of today's acquisition systems record the original content in files, however, which can be instantly used in an editing system without further digitizing being needed. Nevertheless, in some cases, a change in the digital format may be required when uploading acquired content to an editing system, a process typically referred to as *transcoding*.

File Interchange and AAF

As with video servers, the file formats used by different non-linear editing systems are often different, depending on the manufacturers of the particular system. This has complicated the interchange of program material during the program production and postproduction process. A standard known as AAF (for Advanced Authoring Format) makes such interchange easier. AAF is compatible with MXF, and makes provision for the specific types of metadata that accompanies program material during the production and postproduction process.

CHARACTER GENERATORS AND COMPUTER GRAPHICS

Character Generators

A *character generator* is basically a computer system for producing captions, text, and simple graphics for adding to a video program. It may be a regular PC equipped to output video signals and running special software or it may be a dedicated, integrated hardware/software system.

Text input may come from a keyboard or from a communications port. The user can select the background color and the font, format, and color for the text, and how it is displayed on screen (e.g., as rolling titles). In addition to alphanumeric characters, character generators also allow other digital images, such as logos, to be recalled from storage and added to the screen.

All titles and other pages of text that are generated can be saved and stored on a hard drive. They can then be recalled and replayed on demand when needed in a program.

An example of how keyboard input to a character generator might be used would be when a news reporter's name is added to the bottom of the screen during a newscast. An example of an external source supplying data to a character generator through the data port would be when a warning about a weather emergency is received by a television station's Emergency Alert System equipment, and the warning is automatically fed out over the air, using the character generator to scroll a text message across the screen.

Graphics and Paint Systems

Graphics systems may be based on PC platforms, or high-end systems may use dedicated, integrated hardware and software. They provide the capability for producing all types of graphic art, charts, pictures, animations, and so on and may be interfaced to other systems for importing data and other images, such as financial information or weather maps.

Paint systems simulate regular artist painting methods and provide a special user interface to allow full artistic expression. This includes a *graphics tablet* on which the operator "paints" with a pressure-sensitive electronic "brush," producing images on the screen.

ELECTRONIC NEWSROOM

The news production process involves editing of program material from tapes provided by news gathering crews or recorded from remote sources through the ingest system. Stock footage has to be located and made available for editing, and suitable graphics need to be produced to add to the story. At the same time, the scripts for the news stories are prepared separately and brought together with the audio and video program material at the time of transmission.

An *electronic newsroom* brings all of these functions together in one system, using video server–based recording, nonlinear editing, graphics systems, databases, special software, and multiple workstations running special software, all connected by a network. This allows most of the news preparation functions to be integrated, so the people involved can work cooperatively to produce the news stories from their own desktops, while accessing the same material from central servers.

Such systems often provide low-resolution *proxy* video to users on their PCs, for use while preparing the news story. This reduces the load on the network and servers that would occur if everyone had access to the full-resolution version.

For transmission, the electronic newsroom system provides the story scripts for the presenter to read and plays out the high-quality version of the finished news stories at the right times.

SIGNAL DISTRIBUTION

Signal distribution is one of the less glamorous aspects of television engineering—there are no flashy control panels, cameras, recorders, or even pictures to see. Nevertheless, it is one of the most important aspects of station design. The design choices made are fundamental to the way the station works and may be difficult to change at a later date. All of the System Considerations discussed earlier in this chapter affect the equipment used for signal distribution. Audio system and distribution issues were discussed in Chapter 8, and they largely apply also to television systems. Many components make up the overall system, but the main ones include the following.

System Interconnections

The way video signals are transported between different items of equipment, and between different areas in a television facility, is similar whether the signals are analog or digital, SD or HD. Video interconnections within the facility use *coaxial cable* (also known as *coax*). The main difference for digital is that digital video signals, especially HD, with data rates up to 1.485Gbps, require coax with very good high-frequency response to carry the signals without degradation. Even then, HD cable runs are limited to about 1000 feet. SD video, both analog and digital, may run considerably longer than this. For very long runs, up to many miles, video signals may be carried over *fiber-optic* cables.

Compressed bitstreams (see Bitstream Distribution and Splicing, later in this chapter) are also transported within the station on coaxial cable, in much the same way as serial digital video.

Interchange of files between video servers and other equipment uses computer network technologies, with switches, routers, and LAN cabling as discussed in Chapter 6.

Video Patch Panel

Video patch panels serve a similar purpose to the audio patch panels, described in Chapter 8, for testing and cross-connecting program signals. They use a different type of jack, designed for video, with those intended for HD digital video requiring a much better frequency response specification than those for analog signals.

Video Routing Switcher

Most stations have a *video routing switcher* (also known as a *router* or *matrix*) to distribute signals to multiple locations in the facility. Functions are similar to the audio routing switcher described in Chapter 8, but for video, and different versions are needed for analog, SD digital, and HD digital. In many cases, video and audio routing switchers are linked together as separate *levels* of a single unit, to switch both video and audio signals together.

Video Distribution Amplifier

A *video distribution amplifier* (referred to as a DA) has the same function for video as the audio DA described in Chapter 8, for distributing signals to multiple destinations. Video DAs come in analog and digital versions, for both SD and HD. They may also incorporate *cable equalizers* that adjust the frequency response of the amplifier to compensate for the high frequency loss of the cable connected to the amplifier. This is important to maintain video quality when feeding signals over long lengths of cable.

Embedded Audio

To reduce the amount of separate cabling and distribution equipment, some networks and stations use a system based on SMPTE standards for *embedded audio*. This system embeds up to 16 channels (eight pairs) of AES/EBU digital audio signals into the vertical interval data area of SD or HD digital video signals. They can then be carried through most parts of the video system together, without any separate audio switching or processing.

The disadvantage of an embedded audio system is that, whenever it is desired to change the audio (e.g., for editing or mixing new content), it is necessary to disembed the audio and then re-embed it again afterward. There are also problems when video is passed through devices such as DVEs, where, again, the audio has to be disembedded first.

VIDEO TIMING

The *timing* of video signals being mixed or switched is very important, particularly for analog video. If two analog video signals are switched from one to another when their sync pulses are not precisely aligned, this will result in a "bounce" or other disturbance in the output video signal when viewed on a picture monitor or receiver. If two nonsynchronized signals are mixed together, then the resulting picture will be distorted and may become unstable, and if the chrominance signals are not aligned within very small tolerances, color errors will result. Therefore, all signal sources in a studio are synchronized with a signal that comes from a common *sync pulse generator* (SPG), and the lengths of video cables for different sources feeding video switchers are carefully matched to maintain timing accuracy.

Where video signals come from remote sources, which cannot be synchronized, they are passed through a device called a *frame synchronizer*. This is able to store a whole frame of video and automatically adjusts its output timing to synchronize the signal with station video signals.

With digital video systems, the timing accuracy of signals arriving at a switcher is less critical than with analog, because the receiving device can accept digital signals with some degree of timing variations and still produce stable outputs. Nevertheless, all sources in a digital station are synchronized together to avoid timing problems building up, and frame synchronizers are still required for remote sources.

We mentioned earlier that a DVE unit delays a video signal by one complete frame. This generally is acceptable from a video timing point of view because the horizontal and vertical sync points repeat every frame, so the signal can still be switched, mixed, or recorded. It does, however, create a problem for *lip sync*, which we discuss later in the chapter. It may also cause some timing problems in systems still using analog video with color subcarrier.

FILE-BASED WORKFLOWS

Digital production systems for television programs that are not broadcast live have increasingly moved to computer-based operations. This has given rise to a different sort of distribution pathway, in which content is transferred between locations in digital files, using IT-based, computer networking processes rather than the signal distribution and routing methods described earlier. Such *file-based workflows* provide numerous advantages in the speed and efficiency of production, and they enable a wide range of additional creative opportunities. For example, content can be sent from place to place in the facility faster than real time, processes (such as audio loudness management) can be applied to programs in an automated and faster than real-time fashion, and different processes can be applied to the same piece of content or program by multiple production teams throughout the facility simultaneously. For non-live programming, file-based workflows are becoming the norm in television production facilities.

AUDIO FOR TELEVISION

Audio systems and individual items of equipment used for television have many similarities to those used in radio, but there are some differences and additional considerations as described in the following sections.

TV Audio Production

A typical television studio setup often involves many more microphones and live audio sources than for radio. Studio microphones suitable for use in vision are used and, for some types of production, a long *microphone boom* may be used to locate a microphone near the action at high level but out of camera sight.

Wireless microphones are frequently used to allow artists to move around unencumbered by microphone cables. The artist wears a small radio transmitter, or it is built into the microphone itself; a receiving antenna and receiver away from the studio floor picks up the signal and feeds it to the audio mixing console.

The studio audio mixer output is recorded with the video on a videotape machine or server, rather than with an audio recorder. Dedicated audio workstations are not generally used in television, although hard disk audio recorders may be used for some sources and functions, such as sound effects.

Audio Editing and Postproduction

Audio editing for television is usually combined with the video postproduction process. This may be in a video edit suite, using a conventional audio mixer, when sources from VTRs are played back and re-recorded to produce the finished program. Audio editing can also be carried out as part of the nonlinear video editing process, and such systems provide for separately mixing and processing the audio.

Audio Sweetening

The audio that was originally recorded with some programs often needs to be modified and enhanced. This is known as *audio sweetening* and is part of the postproduction process.

Surround Sound

As described in Chapter 16, DTV has the capability to carry surround sound audio. This requires production, network, and station facilities to have audio systems that can handle six channels of audio. In particular, audio mixers, monitoring, switching, and distribution systems, previously intended for stereo audio only, all have to be upgraded with provision for surround sound operations.

Dolby E

Distribution and recording of surround sound signals requires six channels of audio. This creates a problem because many VTRs have only two or four audio channels, and network distribution typically carries one stereo audio signal only. A system called *Dolby E*, developed by Dolby Laboratories, can be used to lightly compress up to eight audio channels into the bandwidth occupied by a single AES3 digital audio signal. This signal can then be recorded onto a single digital VTR or video server, and can be distributed over an AES3 stereo audio channel from a production or postproduction facility to the network, and on to the broadcast station. There it can be converted back to individual audio channels for mixing and switching in master control, and finally re-encoded as AC-3 audio for DTV transmission.

Audio Processing and Loudness Management

The DTV audio system is able to carry a much wider dynamic range of audio than NTSC, so dynamic range compression before encoding is not strictly necessary. In theory, the AC-3 decoder in the DTV receiver is able to carry out audio compression to suit the taste of the listener and the listening environment.

In practice, some, but not all stations do process DTV audio before AC-3 encoding, as shown in Figure 9.11 later in this chapter.

The wide dynamic range provided by DTV audio does present a special problem, however. Television broadcasts typically include content elements originating from different sources, which are assembled sequentially for presentation on a given program channel. Some of these programs can have a narrower dynamic range than others (through the use of heavy dynamic range compression on some segments, making them sound louder than segments without such compression), and this often means that one program segment may sound louder to the listener than the segment immediately preceding it, even though both segments' audio signals are within allowable bounds of the audio system, in terms of digital and electrical levels. Such *loudness mismatches* can be quite annoying to viewers, since they will have set their listening volume to a comfortable level for the first program segment they encounter when tuning to a television channel, and when a much louder segment follows on the same channel, listeners must make quick adjustments of their listening volume downward. When yet another segment follows that returns to a lower loudness level, the viewer has to once again adjust the listening volume upwards, and so on. Often this phenomenon is encountered when a news, sports or entertainment program segment (so-called *long-form* content) breaks for interstitial announcements like commercials or promos (so-called *short-form* content). It is particularly annoying, for example, when a dramatic or suspenseful entertainment program ends one of its segments with quiet dialog or music, and the TV station then runs a commercial announcement that starts with a much louder apparent volume.

It is therefore important to distinguish between audio level and audio loudness. Audio level is what the typical meters on audio equipment display, and operators always attempt to keep these levels in the proper zone of operation that ensures proper operation, and avoids either excessive noise buildup from too low an audio level, or distortion from too high an audio level. Audio loudness, on the other hand, is the perceived volume of an audio signal to a human listener. Two soundtracks with the same, appropriately controlled audio levels can have substantially different loudness, primarily due to one soundtrack's audio level being higher more of the time (perceived as "louder") than another with a more widely varying audio level over time (perceived as "quieter").

When a human operator is involved as a program stream is assembled, this adjustment could be made manually in real time, and often was in earlier times. Alternatively, an audio processing device could add heavy compression to all content segments as they were broadcast (as was typically done in analog TV broadcasting), thereby automatically *normalizing* the loudness between segments. This approach also results in a low dynamic range for the station as a whole, however, which reduces the quality of the audio content and may change to content producer's intent for the program's soundtrack (e.g., explosions are no louder than dialog). The arrival of digital TV, with its possibility for an impressively wide dynamic range for high audio quality coupled with the increased use of automation for program assembly, caused more frequent occurrences of disturbingly wide loudness shifts between program segments on DTV channels.

The CALM Act

Such loudness shifts between program segments are something that the DTV audio system is technically able to accommodate without continual adjustment of the listening volume by television viewers—through a process in the AC-3 audio system called *dialog normalization*—but industry practice did not always take proper advantage of it. This led to many listener complaints, and ultimately engendered a U.S. law called the *CALM Act* (for Commercial Audio Loudness Mitigation), along with subsequent FCC regulations enforcing it. Since 2012, U.S. television broadcasters have thereby been bound by law to maintain a fairly consistent loudness across all program segments.

To assist stations in complying with this requirement, new metering systems have been developed that can quantify not just electrical audio levels but also indicate human perception of long-term loudness, using an internationally standardized loudness-assessment algorithm called *ITU-R BS.1771*. This capability can be used in conjunction with the dialog normalization feature mentioned above to ensure that audio loudness in DTV content remains stable between content segments (in particular, between a television station's programs and its commercials). Further detail on audio loudness management in DTV broadcasting is provided in Chapter 16.

Audio-Video Synchronization

The term *audio-video synchronization* refers to the fact that the audio in a television program must match up—or be synchronized—with the video picture that accompanies it. Synchronization errors are particularly noticeable and objectionable when people are speaking on camera, so they are also known as *lip sync* errors (i.e., when a person's mouth movements do not correspond with the sound of the speech being heard).

Audio-video sync errors may be introduced whenever the audio signal is distributed or processed separately from the video. They may occur in many parts of the program chain, from production through to the receiver, and this applies in both analog and digital systems. There are, however, more ways in which errors may occur when using digital systems, and the errors can be much greater when problems do occur. It is important that such errors are prevented or corrected. If not, even small errors in several parts of the system can add up to significant and objectionable errors by the time they reach the viewer. This is not a trivial engineering task—a lot of effort is needed to maintain synchronization throughout the television distribution chain.

An example of a source of lip sync error is when a video signal is passed through a frame synchronizer. In that case, the video signal will be delayed by at least one frame or about 33 milliseconds. Errors may also be introduced when audio and video signals take different paths on a program link, say for a remote news contribution to a network or station, or when audio and video signals are edited separately in an edit suite or nonlinear editor.

Such errors can be corrected by inserting either audio or video delays into the system, to bring the signals back into synchronization. A delay is a device that takes the audio or video signal as an input and stores it for a period of time before sending it out. The complication is that the synchronization errors are often not constant (e.g., a video signal may sometimes pass through a DVE and sometimes not), so the compensating delay also needs to be adjusted. Even if a television studio is well synchronized internally, it is usually necessary to have additional adjustable delay units available, in case a signal being fed into the studio from outside is out of synchronization and needs to be corrected.

While there are differences of opinion, a commonly accepted rule for broadcasters is that audio signals should not arrive later than about 60 milliseconds after the video to which they correspond, or earlier than about 30 milliseconds before the corresponding video, when seen by the final viewer at home. Outside of these ranges, the errors will be noticeable to many people. The delay can be larger when the audio is later, because that is the natural state of affairs seen in real life, given that light travels much faster than sound through the air. We therefore have higher tolerance for sound being slightly behind the image, as opposed to sound arriving first, which rarely occurs in nature. Because there is some degree of uncertainty in audio-video timing caused by the DTV transmission process, broadcasters need to keep audio-video sync at the input to the ATSC encoding equipment at the studios to much tighter tolerances than the amounts mentioned above, however. The goal is to correct for zero lip sync error at each stage of the production and distribution process.

ANCILLARY SYSTEMS

Television studio requirements for ancillary systems such as clocks, timers, intercom, talkback, and on-air lights are very similar to those described for radio in Chapter 8. Obviously, the scale of such systems is usually larger, to cater for the increased size and complexity of most television facilities.

The station master clock is synchronized with a generator for SMPTE timecode, which is distributed to all equipment that needs a time reference. In this way, both staff and equipment in the station are working with exactly the same accurate time.

Communication systems are another key ancillary system in a television station, as described in the Studio Control Rooms section earlier in this chapter. The signal paths must also be distributed throughout the facility and to remote locations via ancillary communications routing systems.

INGEST AND CONVERSION

Ingest

Program feeds come into a television network or station through various links and routes. These may include satellite and microwave radio links, fiber-optic cables, coaxial cables, and physical videotape delivery. The process of *ingest* includes checking material as it arrives, recording as necessary onto videotape or servers, and routing to the correct destinations for further processing or transmission.

Ingest Automation

Much of the ingest process may be controlled by an automation system. This system keeps track of schedules for incoming programs from satellite links, where material is stored, what format conversion and so forth may be required, machines to use, and other necessary details.

File Transfers

Rather than sending material as audio and video signals in real time, many networks and stations are moving to systems that distribute program material as data files containing video, audio, and metadata (information about the program material). These are transported over private or public data networks, using Internet protocol, hence the process is often referred to as *video over IP*. The incoming files are ultimately stored on video servers. (Such file transfers are utilized throughout the file-based workflows described earlier in this chapter.)

As mentioned in the Video Servers section, the SMPTE MXF specifications standardize the interchange of program material as files, with great advantages for the broadcaster.

Format Conversion

In the United States, programs may be produced in standard definition, having about 480 active lines, or in one of the high definition formats, having 720 lines or 1080 lines. For various reasons, a network or station may wish to use a program in a different format from the one in which it was produced. For example, it may need to use an HD program as part of a standard definition multiplex, or incorporate some SD material as a segment in an HD program. In each case, a format converter is used to take the input signal and convert it to the required format using digital signal processing.

This process is referred to as *upconversion* when the number of lines is increased, and *downconversion* when it is decreased. Upconverted SD pictures can never look as good as true HD material, but good quality can be obtained if the original material comes from high-quality cameras and has passed only through a digital distribution chain.

With format conversion, the picture aspect ratio often has to be changed between 4:3 and 16:9. This requires decisions to be made on whether any of the picture will be cropped in this process, what size black bars (or other colors) will be added to the sides or top and bottom of the picture, and whether any anamorphic squeeze or stretch will be done to help fill the output frame with picture. Sometimes a combination of all these techniques may be used.

Standards Conversion

As mentioned previously in this book, different television standards are used in different regions of the world. When television programs that were produced in accordance with one television standard, such as 625-line PAL, need to be used by a broadcaster with another transmission standard, such as 525-line NTSC, then the conversion is done with a *standards converter*. The process has some similarities to format conversion, but the big difference is that the picture frame rate has to be changed, typically from 25 frames per second to approximately 30 frames per second, or vice versa. Conversion between PAL, SECAM, and NTSC, in any direction, is frequently required for international program exchange of analog video.

Similar conversion may be required between digital video formats, as well, although these are not coupled to the transmission standard as they were in analog broadcasting, and so may vary even within a region (or be the same across different regions, thereby not requiring conversion), based purely on the choice of the individual broadcaster or production facility. Such digital conversions are more appropriately considered format conversions as described immediately above, or may be referred to as *transcoding*.

IP-BASED STUDIO INFRASTRUCTURE

As broadcast facilities increasingly turn to computer networking for interconnection, sections of the facility, or the entire facility, may be based on Internet Protocol (IP) interfaces, using Ethernet or IP routers and CAT-5 or similar cable, rather than dedicated video switchers and coaxial cable. As with IP-based audio facilities, there are numerous advantages in terms of reduction in the number and cost of wiring, ease of installation, and simplified expansion as station needs change over time. Leverage of the economies of scale from the computer industry (as opposed to the specialized equipment produced for professional video applications) can substantially reduce infrastructure hardware costs. There are disadvantages, however, a primary one being the added *latency* (i.e., delays in getting signals through the facility) that may be experienced on certain signal paths, and the possible lack of interoperability between different manufacturers' implementation of IP-based systems.

TELEVISION MASTER CONTROL

Master control contains the equipment for control of the on-air program, whether from a network release center, a station group centralcasting location, or a local station. System architecture for this function varies considerably for different networks and stations, so what follows should be considered as an example only.

Figure 9.11 illustrates in very simple terms how a master control system might be configured for a local DTV station. It shows various local and remote sources feeding a master control switcher. The video and audio outputs from this switcher then feed the ATSC video and audio encoders and multiplexer (audio processing at this point is optional), and the output goes to the transmitter.

Arrangements for a *network master control* may be similar, but the master control switcher output in that case will feed the encoders for the network or group distribution, and then go to a satellite uplink rather than a terrestrial transmitter. Arrangements for a station group *centralcasting master control* will have many of the same elements.

The simplified figure does not show any of the monitoring that is normally associated with a master control facility.

Sources

The principal local sources of video programming and advertisements are most likely one or more video servers. Figure 9.9 shows a large array of servers installed in a major network playout facility with multiple program feeds; local stations will be less elaborate. Other sources may include VTRs, graphics systems, character generators, an audio recorder, and a microphone for live announcements. One or more local studio outputs will also feed the master control switcher.

FIG. 9.9. Server-based Playout System. Courtesy PBS and Omneon Video Networks

Remote sources include the feed from the network and other sources—perhaps a live sporting venue or remote truck. If the station is producing a high definition output, but its local or remote sources are in standard definition, then an upconverter is required for the SD source material. It may also include the feed from a group centralcasting location, which may be left permanently selected during periods when the station service originates at the central location.

There may well be more program sources than there are inputs on the master control switcher, so it frequently works in conjunction with an audio and video routing switcher, to make all the sources available for selection on air.

Master Control Switcher

The master control switcher has some similarities to a production switcher, in that it has input selector buttons, one simple M/E unit, and usually a DVE. It also has key functions for adding graphics and captions to the broadcast program. An example of the control panel for a master control switcher is shown in Figure 9.10.

Differences from a production switcher are that the master control switcher has audio inputs as well as video for each source and is able to switch them

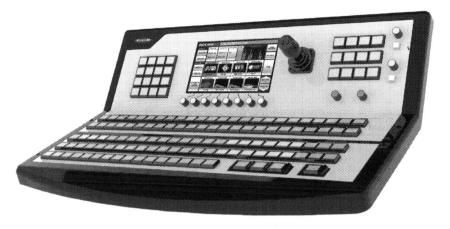

FIG. 9.10. Master Control Switcher. Courtesy Miranda

together, with an *audio-follow-video* function. It can also switch or mix separate audio-only sources for announcements or music. Some master control switchers have functions for starting VTRs or other machines when the source is selected.

When operated manually, a master control switcher works in *preset-take* mode. The operator preselects the next source using a push button on the *Preset* bus. At the time the switch is required, the *Take* button is pressed and the new source is switched to air. Transitions such as wipes and fades can also be executed on cue, together with effects such as *squeezeback*, in which a DVE is used to allow a short segment to be shown on part of the screen simultaneously with the end of the previous program, which is squeezed into a smaller section of the screen (for example, while credits are rolling). Captions and graphics can also be keyed over the program video, including the station logo (often displayed in the lower left corner of the screen, and sometimes called a *"bug"*) or Emergency Alert System (EAS) text.

Audio announcements, either live or prerecorded, may be added using a *voice-over* facility, which drops the level of the main program to allow the announcement to be heard. The switcher may also make provision for inserting EAS audio in place of the program audio. There are outputs for feeding picture monitors for the preset and program signals, and also for monitoring the audio. Audio level metering may be built in or provided externally.

This type of equipment is available to switch analog, SD digital, or HD digital signals. Prior to completion of the DTV transition, when stations operated two separate outputs (one for NTSC and one for DTV), there may have been two master control areas. Alternatively, one master control position may be able to control two switchers from the same control panel. Similarly, when a station operates several DTV program services for *multicasting* (see Chapter 16), the different outputs may be switched from one control panel, again controlling several switchers.

Emergency Alert System Equipment

As noted in Chapter 8 on radio studios, EAS equipment must be installed somewhere in a broadcast station's air chain. In a radio station, this equipment is installed between the output of the on-air mixing board and the feed to the transmitter. In a local television station, the arrangements are similar but, in this case, the equipment must have the ability to interrupt both audio and video, or at least to interrupt the audio and insert a video text message over a portion of the normal picture. One arrangement for this is shown in Figure 9.11 below. Here, the EAS audio output feeds the switcher, which switches that source to the output on demand. The EAS text output is fed to a character generator, which generates the video text, and this is keyed over part of the video screen. In an alternative arrangement, the EAS announcement switching may be carried out downstream from the master control switcher.

Time Delay

Master control for a network may feed several network distributions intended for local stations in different time zones. For the U.S., assuming the live network feed originates in the Eastern time zone, the others can be produced just by inserting a suitable delay. Traditionally, this was done by recording the output on a VTR and playing back the tape at a later time (known as *tape delay*). The current method is to use a video server that records the live network feed and provides multiple outputs of the same material with different time delays. It can start replaying the delayed material while still recording the current program.

Master control in both network and station output feeds may need a smaller, switchable delay of several seconds to allow for deletion of profanity or undesirable content from live shows before it is sent to air. Although this task can be performed with a disk-based server, digital delay units are available for just this purpose, using solid-state memory.

TELEVISION AUTOMATION

On-Air Automation

Manual control of a master control switcher is possible for simple operations, but it gets very difficult for an operator to implement current network or station playout schedules. These are often a complex sequence of many segments for programs, advertisements, promos, and other material that need to be switched on air at the right time, often in quick succession. The problems are compounded with a multitude of graphics, captions, and special effects to be controlled, and multiplied again if there is more than one output channel.

The use of automation to support the on-air master control process is, therefore, almost universal today. It reduces on-air mistakes and ensures that advertisements or other interstitial messages are played out as scheduled, which is important since the airing of such announcements is a primary source of revenue for most television stations. The on-air automation system controls the local servers, VTRs, and other sources. It finds and cues program material using SMPTE timecode as the locator. It starts machines and controls the switcher that puts the source on air, initiates transitions and effects, and keys captions and graphics.

Playlist

The key to automation control is the *playlist*. This is based on the station program schedule and includes all the *events* that need to take place to produce the continuous program output from the station. Each event has a time entry indicating when it needs to take place. At the correct time, the automation system sends commands to the appropriate equipment (e.g., "Start" to a VTR) to initiate the action needed for the event.

The playlist is generated some time before transmission, taking input from the station *traffic* system, which is responsible for keeping track of what advertising has been sold, when it will be played out, and what all the segments are for the program schedule.

Other Automation Functions

It was mentioned previously that the process of ingest may be supported with an automation system. Automation may also be used to support other aspects of media preparation, keeping track of what material is available, and where it is stored. The electronic newsroom system also uses automation, both for content ingest and playout.

Integrating these systems with the on-air playout system, or at least having links to allow them to share information, minimizes the need for manual intervention and reduces the likelihood of errors throughout the system.

With some station groups, there may be a centralcasting function that involves centralized management of the station automation system, with some content being downloaded to the station and played out from a local server under automation control, rather than complete assembly of the program happening at the central location. Specific arrangements may vary considerably among different stations.

ATSC ENCODING AND MULTIPLEXING

Chapter 16 explains how digital video and audio has to be compressed and combined into a single *transport stream* before it can be transmitted. This is necessary because the data rates of the video and audio signals that come from the video server or other sources are far higher than can be transmitted within the broadcast channel's bandwidth.

As shown in Figure 9.11, the audio and video compression is carried out in an *ATSC encoder*. The encoder produces two data streams (one for audio and one for video) called *elementary streams,* which are combined into a single stream of data packets in a *multiplexer* (or *"mux"*). The output of this multiplexer is then combined, in a second multiplexer, with other packets of information (see following sections) to produce the final *transport stream* that can be fed to the transmitter.

Depending on the equipment design, the multiplexers may be separate devices or may be in a single integrated box with the video encoder. The audio encoder may be combined in the integrated unit or may be a stand-alone device.

Most stations locate the ATSC encoding equipment at the studio center and feed the single transport stream over the *studio-transmitter link* (STL) to the transmitter.

MULTICASTING OPERATIONS

Figure 9.11 shows the basic station output arrangement for a single DTV program service. Many stations multicast two or more program services in a single output bitstream on their DTV channel, however. In that case, there will be multiple video and audio feeds from master control, and multiple video and audio encoders. These multiple program services are all combined in the final mux, as indicated in the figure.

CLOSED CAPTIONING

The FCC requires most broadcast television programming to be transmitted with *closed captioning*. This allows a viewer with a hearing disability to see a text version of the spoken audio on the screen, when the program is displayed on any television set with its closed captioning feature turned on by the viewer.

Most television programs have their captions inserted into video content by third parties that specialize in the service, using either human operators or automatic *speech-to-text* ("voice recognition") software. Programs received by a station from a network or other supplier typically arrive with caption information already inserted in the video signal. Live local programs, such as news, have to be captioned at the station, although once again, the captions may actually be generated off-site and inserted as the content is broadcast. A small delay between the live audio and the corresponding caption text is permissible.

In the NTSC era, the closed captioning data was carried in line 21 of the *vertical blanking interval* (VBI) of the video signal, as described in Chapter 15. When this video was broadcast, the VBI data was simply carried through the NTSC transmission chain. In DTV, the situation is more complex. Caption data embedded in video program content must be extracted by the DTV station and placed in the *User Bits* part of the ATSC encoded *video bitstream*. This process may take place entirely inside the ATSC encoder, or may require an external device, called a *caption server*, as shown in Figure 9.11. If the station is generating its own captions (e.g., for a local newscast), a caption server is also required.

VIDEO DESCRIPTION

The FCC also requires a certain amount of television content to include *video description*, which is an additional audio track that describes the scene on screen for visually impaired listeners. It is added to the regular soundtrack of the program in such a way that it does not interfere with the program's dialog. Video description is presented as a second audio service, typically mixed with the program's original soundtrack, which can be selected by visually impaired viewers instead of the regular soundtrack. Again, these alternate soundtracks are generally produced by third parties, but unlike closed captions, producing the service is not simply a text-generation exercise. The video description scripts must be written, recorded and appropriately mixed with the original soundtrack so dialog is not interfered with. Once again, the video description audio service is encoded and multiplexed into the ATSC transport stream for broadcast.

ALTERNATE LANGUAGE AUDIO

In some cases, one or more alternate language soundtracks may be provided for a video program. As with video description, the alternate soundtrack(s) is/are encoded and multiplexed with the ATSC transport stream, and can be selected by the user in lieu of the main audio soundtrack. Unlike video description, however, the alternate language soundtrack(s) is/are completely independent of the main program soundtrack, and can therefore be produced more simply, even live and in real time (e.g., from a second announce booth on a sports broadcast).

PSIP GENERATOR

Program and System Information Protocol (PSIP, see Chapter 16) tables are produced in a computer called a *PSIP generator*. The output of this is a data stream, which provides program schedule and description information to viewers, feeds into the ATSC multiplexer or ATSC encoder (if they are combined together in one unit), as shown in Figure 9.11. The PSIP generator is usually connected through a network with other computers, such as the *traffic system*, and *automation system*, which produce and manage the program schedule and associated information. This information, known as *metadata*, is needed to help generate the PSIP tables. An ATSC standard called *Programming Metadata Communication Protocol* (PMCP) enables it to be shared among different systems and equipment.

DATA BROADCASTING EQUIPMENT

For DTV stations, a *datacasting* service may be combined with the television programs in the DTV *multiplexer* at the output of the studios, as shown in Figure 9.11. This requires a data connection to the ATSC multiplexer or ATSC encoder (if they are combined together in one unit) from the *data server*. The data server stores and communicates the data service and is usually connected through a network with other computers, either at the station or elsewhere, which produce and manage the data information.

ADVANCED PROGRAMMING SERVICES

Besides providing "pure" data delivery services (i.e., unrelated to broadcast content) for specialized purposes and individual clients—such as a company or a school district wishing to distribute data to all its branch locations in a metro area—DTV data broadcasting can also carry advanced television program services intended for the station's broadcast audience, which can be developed and added over time.

One such extension is *Mobile DTV*, also known as *ATSC Mobile/Handheld* (ATSC M/H). This service acts as a datacast in the ATSC multiplex, and carries highly compressed audio and video with additional error correction to make it adequately robust for mobile and handheld reception to M/H receivers. (Standard ATSC digital TV services are generally only receivable by fixed receivers.) Like a multicast channel, broadcasters can program the M/H service separately from the fixed services on the multiplex, and multiple M/H services can be implemented on the multiplex, up to the limit of capacity in the broadcast channel. Audio-only M/H services are also possible. Therefore studio facilities must be able to generate and encode additional video and audio channels for M/H services.

Another type of service that can be added to an ATSC DTV multiplex is Non Real-Time (NRT) content. This is a data service that provides a variety of options for broadcasters to present on-demand type programming to viewers over the air. In this case, station studio facilities require servers and special encoders to distribute content in file-based form to NRT-capable receivers.

Yet another type of possible DTV content offering is a type of ATSC data broadcasting that allows viewers to interact with the content, typically via an Internet *backchannel* or return path. This class of services is generally referred to as *ATSC 2.0*, and it provides methods by which viewers can personalize or enhance broadcast content at their option via an Internet-connected ATSC 2.0 receiver. To provide this kind of service, broadcast facilities need servers and adequate connectivity bandwidth to provide the additional content and respond to viewers' requests. Alternatively, a broadcaster may outsource some or all such interactive responses to one or more third parties.

Finally, DTV broadcasters may be able to offer 3D television services, through a variety of different ATSC methods that may involve data broadcast services. In such cases, broadcast studios require appropriate equipment for the generation, encoding and monitoring of 3D content.

More details on advanced ATSC services can be found in Chapter 16.

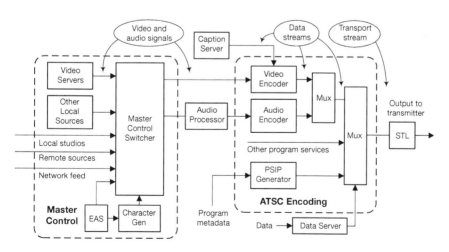

FIG. 9.11. DTV Master Control and Emission Encoding

BITSTREAM DISTRIBUTION AND SPLICING

The previous sections on master control and ATSC encoding for a DTV service assume that the program feed from the network is converted from its incoming distribution format to uncompressed baseband video and audio before being routed around the station, switched to air through a master control switcher, and then ATSC encoded for transmission.

Not all networks and stations follow this model, however. In an alternative system architecture, the network carries out the ATSC encoding for the network DTV program feed (usually, but not necessarily, in high definition) at the network release center. It then sends the much lower data rate, compressed transport stream through the network distribution (usually via a satellite link) to local stations. This produces considerable savings in satellite circuit costs and helps maintain quality by reducing the number of times the network signal has to be encoded and decoded. This network feed is effectively ready for transmission at the station, after it has local PSIP information added.

With this arrangement, the local station prepares its own local material as previously described and encodes it with its own ATSC encoder. However, whenever a switch is required from local to network material, this is done by means of a *bitstream splicer*, which switches seamlessly between the local and network bitstreams. Special arrangements are made for inserting the local station logo on the network feed without decoding the compressed stream.

It is possible for control of the bitstream splicer to be integrated into a master control switcher for the local sources, and controlled from the station automation, so there is no overall increase in operational complexity.

Compressed Bitstream Distribution

Compressed bitstreams are distributed within the station using similar techniques to serial digital video. There are two main standards for this: SMPTE 310M, which is used mainly for the connection from the multiplexer to the STL, and DVB-ASI (Digital Video Broadcasting Asynchronous Serial Interface), which is usually used everywhere else.

Bitstream Recording

The ATSC-encoded transport stream can be recorded on a *bitstream server*. Thus, the ATSC-encoded feed from a network may be time-shifted for transmission by the station at some later time than when it is distributed to the station by the network. This is a very efficient method of recording fully prepared HD signals for transmission, but it is not very practical for production or postproduction.

Mezzanine-Level Distribution

In addition to the "uncompressed" baseband or "fully compressed and ready for ATSC transmission" architectures, there is a third alternative, which exists in between the other two methods. It is based on an intermediate, more lightly compressed bitstream, sometimes known as *mezzanine level*, typically with a data rate of 45 Mbps, which is the rate that many networks use for their DTV network distribution. Many servers also record a lightly compressed bitstream, using various formats. Stations typically decode these signals back to uncompressed baseband video and audio for distribution in the station, but it is possible to distribute the mezzanine-level signals in the station.

SIGNAL DELIVERY TO MVPD HEADENDS

Most U.S. television viewers do not receive local broadcast stations' signals directly off-the-air via an antenna, but instead access them as part of a package of television channels delivered via a pay-TV (cable or DBS) service. These services' programming packages usually include all the local broadcast channels available in the viewer's geographic area, along with many other non-broadcast (and generally non-localized) channels. Such services are generically referred to as Multichannel Video Program Delivery (MVPD) systems.

Traditionally, MVPDs received local broadcast station's signals over the air via an antenna, and inserted them into their service packages. Increasingly, however, the MVPD may arrange with a station to receive its signal via a fixed link (e.g., a wired or fiber-optic data path provided by a telecommunications service company). Such a signal path would run from the station's studio facility to the MVPD's *headend* (the facility where it assembles and prepares its multiple channels of video programming for distribution to its customer). In some cases, a station may have several such additional feeds from its studios, to serve each of the several cable systems that may each serve different portions of the metropolitan area that the station covers, and/or to feed one or more DBS services.

These are generally wide-bandwidth links that may provide a higher quality and/or more consistent signal from the broadcast station to the MVPD. These links may also allow additional content or services to be shared between the broadcast and MVPD facilities.

INTERNET TV SERVICES

When a broadcaster wishes to create a stream of the station's video or audio content for Internet distribution, the stream may be produced by the station, using a suitable streaming encoder. Alternatively, the broadcaster may provide a regular audio and video program feed to a third-party organization that provides all the facilities and management needed for streaming encoding, together with the Web servers and interface to the Internet.

In either case, the station may use a feed from the same master switcher that feeds the broadcast transmitter, or may perhaps generate a second independent feed, which allows for dropping or adding Internet-specific programming or advertisements.

A station may also offer on-demand Internet service, by which individual, locally produced programs (such as local news broadcasts) may be made available after they are broadcast for later online viewing via streaming or downloading by users at their leisure. These content elements often have additional announcements and/or advertising appended to them, which may differ from those included with the content when broadcast. The station may assemble these on-demand offerings and provide them to online viewers from the station's own servers, or external third parties may be contracted by the station to provide these services. More detailed information regarding Internet-delivered television can be found in Chapter 17.

CHAPTER 10

Remote Broadcasting

Remote broadcasting includes production or acquisition of program material, originating at locations away from the studio center. Broadcasting relies greatly on production that takes place outside the studio, for news, sports, music events, and other content.

Systems for remote acquisition and production activities are described in this chapter, while the link systems for getting them back to the studio are covered in Chapter 11.

RADIO NEWS GATHERING

Radio *news gathering* (content acquisition) includes reporting from the field, recording interviews, press conferences, and so on, for subsequent transport back to studio centers and incorporation into news programs. Most equipment for radio news acquisition uses the same principles as fixed radio studio equipment, although emphasis is on portability, ruggedness, and ease of use in difficult conditions. Microphones used in remote locations are often highly directional with windscreens to reduce the effect of wind noise on the microphone diaphragm. Handheld, omnidirectional microphones may also be used by reporters for "stand-up" on camera reports or interviews, where the microphone can be placed close to the sound source. Omnidirectional microphones' reduced sensitivity to wind and handling noise is a helpful attribute in the field.

Recorders are often small enough to be slung over the shoulder or held in the hand, and while traditional systems used cassette tape or other moving media, most field recording today is performed on devices using solid-state flash memory, either as on-board RAM in the recorder, or more commonly, to one or more formats of commonly available solid-state memory cards. These units usually include both microphone and line-level inputs, and often add built-in microphones, as well. An example of a solid-state flash memory recorder is shown in Figure 10.1. Recorders may have two or more inputs, or a small external mixer

FIG. 10.1. Portable Solid-State Audio Recorder. Courtesy of Tascam

may be used for multiple sources. Monitoring is performed on headphones, and battery power is essential for all equipment.

When covering press conferences, a common audio feed is often provided from the podium microphone(s), which is split and distributed for use by multiple reporters. This is often called a *mult* or a *pool feed*. When this is not available, the field reporter can use a small clamp-on mount to attach a microphone to the podium or any other convenient object. Wireless microphones may also be used for this or other newsgathering purposes. Wireless microphones are especially helpful when the subject and/or the reporter are on the move during recording. In such cases, the wireless microphone's receiver is plugged into the reporter's recorder.

Recorded media may be transported back to the studio for editing, or live contributions may be sent over contribution links of various types, as described in Chapter 11. When on remote assignments, some reporters may take a portable mixer and more than one recorder to allow them to edit a report in the field, before sending it to the station. It is also possible to transfer audio files from a digital recorder to a laptop computer and edit the material on location. At the other extreme, the simplest live arrangement is the reporter speaking over a regular telephone line or cell phone (sometimes referred to as a *phoner*). A similar option is the use of a smartphone app that incorporates an audio codec to send compressed audio from the smartphone microphone back to the station via the Internet or dedicated IP link, which can often provide higher quality audio than a phone call.

Cue Audio

Wherever possible, the audio sent to a remote contribution site for cue purposes is a *mix-minus* feed from the studio audio mixer (i.e., the output of the studio, minus the contribution audio), plus communication cues (see section on Communications Systems in Chapter 9). This helps avoid problems with audio feedback and echo, and makes two-way communication with presenters and talent possible.

The reporter may also have a portable receiver for listening to the station output off-air. This gives the reporter an indication of what is going out over the air from the remote location, but preferably should not be used for cue purposes due to the delay that might be introduced in the broadcast air chain. (This is especially important for stations using IBOC broadcasting, where five seconds or more of delay may be added in the transmission encoding process.)

RADIO REMOTE PRODUCTION

Radio remote productions include live or recorded programming such as sporting events, concerts, promotional events at county fairs and other shows from outside the studio. Facilities required may range from simple to complex.

Equipment is usually transported in packing cases carried in a car or van, and may include portable or smaller versions of equipment used in the studio of the following types:

- Microphones, headphones, headsets (combined mic and headphone)
- Audio mixer

- CD players
- Solid-state recorders
- Monitor loudspeakers

Other equipment that may be needed for some types of remote events includes the following:

- Wireless microphones
- Public address amplifier and speakers (or interface hardware to the house system)

For live broadcasts, some of the same equipment as listed for news reporting may be used for communicating with the studio, and in addition, the following may be needed:

- Remote pickup (RPU) radio link system, or codec for wired digital link
- Portable antenna system for RPU
- Monitoring and communications equipment

More detail on the latter items can be found in Chapter 11.

TELEVISION NEWS GATHERING

Electronic news gathering (ENG) forms an important part of broadcast station operations. As with radio, most of the equipment uses essentially the same principles as fixed television studio equipment. For television, however, the emphasis is even more on portability, ruggedness, and ease of use in difficult conditions, compared to equivalent TV studio equipment.

ENG news crews and their equipment may be transported in a van or small truck, or they may work with a larger truck that includes a microwave link or a satellite uplink for live communications with the studio. In some case, a large production truck may also be used for switching video and mixing audio from multiple cameras and microphones at the event (e.g., a sports broadcast).

Cameras and Recorders

ENG cameras all have built-in recorders, so the term *camcorder* is often used. Most of the small VTR formats have been used at one time or another. Principal formats used have included BetacamSP, DVCPRO, DVCPRO50, and DV, as well as more recent optical-disk, hard-disk and solid-state recorders intended primarily for ENG and *electronic field production* (EFP—see below) applications. Figure 10.2 shows an example of a portable camera with a built-in solid-state recorder.

FIG. 10.2.   Portable Camcorder Using Solid-State Memory. Courtesy of Panasonic

At this writing, many local TV stations produce news programs in high definition, and have equipped their news crews with HD camcorders. Others remain with SD for local production, but most of these stations will likely convert eventually to HD as the cost of HD equipment continues to fall, and SD equipment becomes less available.

Cameras are usually handheld (actually shoulder-mounted) but may be used on a tripod, with a pan-and-tilt head, or installed in a vehicle or helicopter, where a gyroscopic mount is required to produce stable pictures. The crew may have a portable lighting kit to supplement available light where necessary.

Recorded media may be physically transported back to the studio for editing. Alternatively, live contributions or recordings may be sent to the studio over contribution links of various types (see Chapter 11).

Cue Feed

For live contributions, a cue audio and (in most cases) video feed from the studio back to the remote site is required for the reporter to know when to speak and to converse with studio talent. Some ENG links (see Chapter 11) provide a return path via their intrinsic bidirectional nature, while in other cases the ENG link is one-way only (from the remote site to the studio), and communications back to the remote site from the studio are established via separate paths using a variety of cueing links, mobile radios or cell phones.

Field Editing

Ultra-compact editing facilities are available and may be taken on assignment to edit and assemble stories while in the field. Dedicated portable editors were used for this purpose in the past, but increasingly this is performed on portable or high-end laptop computers, using professional video production software.

ENG trucks equipped with microwave or satellite links may also be equipped with basic editing equipment to allow material to be reviewed before sending to the studio.

TELEVISION REMOTE PRODUCTION

Remote field production falls into two main categories: (1) electronic field production using portable equipment and (2) mobile production relying on equipment installed in mobile vehicles.

Electronic Field Production

EFP normally uses portable, self-contained equipment that can be transported in rugged shipping cases. Sometimes, though, EFP equipment may be installed in a small truck or van, so the distinction between EFP and small-scale mobile production is blurred. EFP is usually done with *single-camera* shooting, however, although it is possible to use multiple cameras when required.

Therefore, the main distinctions between ENG and EFP are that EFP equipment and crew may be somewhat larger, the EFP program output is usually recorded on-site for later postproduction, and wherever possible, studio-quality video and audio is maintained in EFP. In addition, sound is usually mixed with a portable mixer before recording on the camera, and mains power may be available, so a wider variety of equipment can often be used for EFP.

Cameras

There is some overlap between cameras used for EFP and high-end ENG cameras. EFP cameras usually have superior picture quality and are typically HD-capable. It is also common for EFP to use a camera with a base station and separate recorder, whereas ENG almost always combines the functions with a camcorder. The separate recording process in EFP allows a separate video engineer to monitor and control picture quality, using a picture and waveform monitor during shooting, while the camera operator concentrates only on the photographic aspects.

Where steady shots are required from a handheld camera that has to move, the camera is fixed to a special counterbalanced and stabilized mount that is strapped to the operator, who can then walk or even run without affecting the picture stability.

Audio

Another distinction involves audio. ENG often uses a microphone on the camera, and/or a microphone that is handheld by the reporter. EFP often uses separate microphones (on boom poles, lavaliere microphones clipped onto talent, or mounted elsewhere around the location), which are handled by one or more audio operator(s).

Mobile Production

Mobile production trucks, also known as *remote trucks* or *outside broadcast* (OB) *vehicles*, have been used for television broadcasting for decades. They provide most of the facilities of a fixed production studio, but in a mobile environment, and can be used for anything from field recording to live remote inserts into studio programs, to complete live-program production from a remote location. From the outside, they generally appear like normal trailer or tractor-trailer trucks. Inside, the trailer unit is typically divided into multiple, air-conditioned areas for production control, audio control, video control, recording, and other equipment. Trucks may be powered from a mains power supply at the remote site, but where this is not available, a trailer-mounted and soundproofed electrical generator may be used. Some trucks have their own built-in generator.

Equipment

Mobile trucks may carry full-size studio cameras or portable cameras, or often both for flexibility in use. Camera mounts are usually tripods. Typically, the cameras will have lenses with very long zoom ranges, to bring action up close, and will have long camera cables, sometimes several hundred meters. Fiber-optic cables are used for distances up to several thousands of meters, or cameras may have radio links, with no cable at all.

Audio is typically picked up by fixed and wireless microphones distributed where necessary on site, and in most cases ultimately sent to the truck via a single, large multi-channel cable (sometimes called a "snake"), for mixing on a production-studio style audio mixing console.

Trucks have storage lockers for the portable equipment in the lower skirts of the vehicle. Cables are stored on drums, which may be motorized for ease of handling. Cable connections for cameras, microphones, and other feeds in and out of the truck are made on a special connection panel in a compartment that is shielded from the weather.

Other equipment types in the vehicle, including switchers, monitors, VTRs, servers, communications systems, and so on, are all much the same as those used in studios.

Slow-Motion Cameras

Special cameras are available to help produce high-quality slow-motion pictures, especially for action replays in sports programming. Slow-motion playback requires more frames than are recorded for a normal speed motion. Therefore, when a conventional camera is used, frames have to be repeated upon playback, resulting in jerky motion. A slow-motion camera works at a higher frame rate, and more frames can be recorded on an associated hard disk recorder. This enables much smoother motion when played back at slower speeds. A special controller is used for slow-motion replay, which provides rapid access to particular parts of the action and controls the playback speed and other functions.

CHAPTER 11

Links

Today the world is a highly connected place, with many options for communications by either wired or wireless means. Yet broadcast links—the communications paths used by broadcasters for the production and delivery of their broadcast services—retain two special requirements that still place them in a special category. These are a need to provide the highest possible audio and video quality from anywhere in the world (or beyond) to broadcast audiences, and a requirement for extremely high reliability.

There was a time when such quality and robustness was hard to come by, and it was only available to broadcasters from a very few telecommunications service providers—often at a steep price. In some cases, broadcasters "rolled their own" paths by setting up point-to-point transmission links, but these were only possible in a minority of cases, and usually only for relatively short, local paths. Subsequently, however, many more options have emerged (although a few have come and gone in the interim), and broadcasters now generally have a wider range of options from which to choose. Nevertheless, as you will see below, some links provided by broadcasters for their own use remain in operation, and are still a viable choice in certain cases. In other cases, a variety of third-party offerings are used by broadcasters for these purposes.

LINK ARCHITECTURES

Broadcast links fall into three main categories as follows:

- *Contribution links* for sending news reports, sports coverage, and other remote production material to a network or station studio
- *Distribution links* for sending finished programs to multiple locations around the country
- *Studio-transmitter links* (STLs) for connecting a local station studio center to its transmitter site(s)

Links can be *unidirectional* or *bidirectional*, and they can be configured as *point-to-point* or *point-to-multipoint* paths. Several types of links are used for radio and television, based on different technologies.

There are a few other parameters of design for these links. They can either be established as dedicated, point-to-point paths (for frequently used remote locations to communicate with a studio, for example), or they can be "user-switched" services. The best known example of the latter is the regular telephone system, officially known as the *Public Switched Telephone Network* (PSTN), but commonly referred to by broadcasters as "POTS", for "Plain Old Telephone Service," or simply "dial-up." Used as a kind of last resort for broadcast contribution, it allows a broadcaster to obtain audio from nearly any inhabited area on Earth, without intervention of the service provider. In other words, once POTS service is provided to a broadcast facility, like any user, a broadcaster can connect it to any other location with POTS service installed, via dialing the number associated with the location. This familiar action of placing a phone call is technically a user-switched service, and the model is applied in other services, as well, like ISDN (see below). In contrast, dedicated paths are "hard-wired" between two locations, and are "always on," without no requirement or option to "dial up" a switched path.

Switched services are themselves provided in two forms. Traditionally (in the analog era), these were *circuit switched* systems, meaning that when a user dialed a number, a connection was established between the wires that served each location, via one or more *crosspoint switchers* managed by the service provider, and that path remained in place for the duration of the call. When digital systems emerged, circuit switched systems were largely replaced by more efficient *packet switched* service, by which small packets of information are each addressed to a specific destination, and they can share physical paths with other packets ultimately bound elsewhere. This approach makes much better use of installed communications infrastructure, and is the basis of the *Internet Protocol* (IP), which is increasingly used for all types of point-to-point communications, including voice calls.

Another parameter involves whether the connection from the user's *terminal* or *point of access* to the communications network is wired or wireless. This generally refers only to one or both edges of the communications path—the so-called *first-* or *last-mile* connections, since the points in between are typically carried on wired paths. Thus, wireless connections are usually short extensions along the edges of a wired network.

A final distinction for packet-switched networks is whether the network architecture is small or large, or in other words, limited to a *local area network* (LAN), or distributed over a *wide area network* (WAN). Examples of LANs are the private computer networks used in homes or offices, while WANs are typically operated by commercial telephone companies or other communications service providers. Both LANs and WANs can include wired and wireless connections to users' terminals. An example of wireless LAN is the familiar WiFi technology, while wireless WANs include *broadband wireless* or "cellular" digital mobile phone services.

Broadcasters use all of the above architectures as appropriate, including a variety of specialized, business-grade links, coupled with use of the same communications systems available to the public, when necessary.

Analog, Digital and IP-Based Systems

Regardless of what path the signal takes for the link (e.g., simple cable, phone line, leased phone line, radio link, microwave link, satellite link), one other important characteristic must be considered—whether the link carries analog or digital signals.

The trend in broadcasting is toward digital signals throughout the chain, but the choice of codec technology and system will depend very much on the signal formats of the audio and video signals that are to feed the link and that are needed for the equipment at both ends. If necessary, *analog-to-digital* (A/D) and *digital-to-analog* (D/A) *converters* may be used to make the interfaces to the link work properly.

Compressed Links

One important consideration for digital links is the data rate of the signal to be carried and the bandwidth that the link can handle. In some cases, audio and video data compression is used in conjunction with the link to reduce the data rate (see Chapters 5 and 16 for more on data compression). When the link is provided to a broadcaster by a third party, the cost of the service will typically vary proportionally with the bandwidth of the link, so broadcasters may opt to use compression for financial reasons. As mentioned elsewhere, the compression systems used for these purposes are often referred to as "codecs."

IP-Based Links

Increasingly, digital links are moving from a variety of specialized and proprietary data formats to the global standard used by the Internet, called *Internet Protocol* (IP). It should be noted that the IP format is also used on many links that are not part of the public Internet per se, so IP links are a common method of both public and private, point-to-point and point-to-multipoint communications, in both LAN and WAN architectures, wired and wireless, throughout the world.

CONTRIBUTION LINKS FOR RADIO

Telephone Line

As noted above, standard PSTN lines ("POTS") or cellular phones are the default links used for radio news reporting. In many cases, the phone is used directly by the reporter for voice transmissions. When the reporter wishes to file audio other than his or her own voice, however, or if the reporter wants to provide somewhat higher audio quality than the phone handset's (or cell phone's) microphone is capable of, various interface devices are available for using outboard microphones and field recorders with the phone line. Figure 11.1 shows a simple analog audio mixer combined with a telephone interface unit. This unit works with a standard dial-up phone line to connect various audio sources at a remote location to the studio, and provides reasonable quality mono audio with a return audio cue feed from the phone line to the reporter's headphones.

FIG. 11.1.   Audio Mixer/Telephone Interface. Courtesy of JK Audio

(Similar interface devices are designed specifically to work with cell phones.) Note that these devices may not work with certain office phone systems (called "PBXs," for *Private Branch eXchanges*).

More sophisticated devices are available that process the audio in special ways to improve the sound quality transmitted through phone lines. Some of them use more than one phone line to provide more bandwidth.

ISDN Line

Integrated Services Digital Network (ISDN) is a telephone-company provided service using switched digital phone lines. Prior to the emergence of the Internet, ISDN was envisioned by telephone companies to be the replacement for the analog POTS service with 64 kbps switched digital lines (called *DS0s*), but this plan never took root. Nevertheless, ISDN was strongly adopted by business users during the 1980s and later, including broadcasters, in both its *Basic Rate Interface* (BRI, providing up to 128 kbps connectivity on two switched DS0 circuits per installation), and its Primary Rate Interface (PRI, providing up to ~1.5 Mbps connectivity on 24 switched DS0 circuits per installation in N. American and Japan, or ~2.0 Mbps on 31 DS0 circuits in Europe, Australia and elsewhere). The system is quite robust and flexible, allowing multiple DS0 circuits to be "bonded" together for aggregating bandwidth into a single virtual link between two given locations.

Using a special codec and *terminal adapter* at each end of the line, full-range broadcast-quality stereo audio can be carried via ISDN BRI, using one or both of its DS0 circuits. While ISDN has become a staple for live radio links, at this writing it is being phased out by most telephone companies, and may eventually become obsolete. In the meantime, however, it remains in use by some radio broadcasters, who value its superior reliability to most IP-based alternatives.

Remote Pickup

Remote pickup units (RPUs) use radio transmissions with various frequencies allocated in the HF, VHF, and UHF bands. Unlike telephone-company provided services, these wireless, point-to-point links are established and operated by broadcasters themselves, under license from the FCC (and often with coordination of a local authority to avoid interference among broadcasters). They can provide reasonably good-quality contribution links without needing a phone line. An RPU has a transmitter with a directional antenna at the remote site—usually mounted on a van or truck, although it can be a portable unit. It sends audio back to the studio, where an antenna is usually mounted on the roof or tower at the studio, with the receiver in the master control area.

T1 Line

A T1 line is a bidirectional, dedicated point-to-point digital data circuit with a bandwidth of 1.544Mbps. It may be leased from a telecommunications service provider or implemented as part of a broadband microwave system. Full-range broadcast-quality stereo audio can be transmitted using special codecs at each end of the line, with full-quality feeds in both directions. It is extremely robust, and fairly flexible in what it can carry or interconnect between two fixed locations.

LAN/WAN

Program contributions may be received as audio files distributed via IP over a computer network. The network may be a virtual private network (VPN) or other designated path established over the public Internet, or it may be set up on private data circuits leased from a telecommunications service provider. Wired LAN or WAN interfaces are typically established on Ethernet connections, while wireless LAN connections use WiFi, and wireless WAN connectivity is typically provided via 3G or 4G broadband wireless service. A wide variety of interfacing equipment is available for radio broadcasters to use in such applications, which may incorporate multiple audio inputs for microphone- and line-level signals, mixing and monitoring capabilities, codec(s), and terminal equipment for one or more of the connectivity options discussed above, into a single, roadworthy hardware package. See Figure 11.2 for an example.

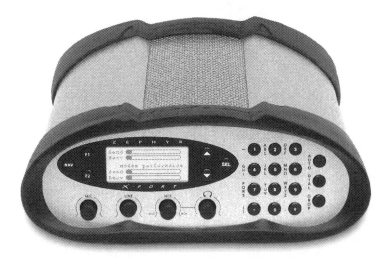

FIG 11.2.   Audio Mixer/Codec/Interface for Analog Telephone and ISDN Lines.
Courtesy of Telos Alliance

CONTRIBUTION LINKS FOR TELEVISION

Microwave

The most common television contribution links are broadcaster-provided (licensed), point-to-point, wireless systems that use microwave radios, with frequencies allocated in the high UHF and SHF bands. Microwaves have extremely short wavelengths that must have line-of-sight transmission. So the transmitting antenna, with a parabolic-shape reflector, is raised as high as possible on a telescopic mast, to get line of sight to a receive antenna mounted as high as possible at the studio center. A station may set up a central ENG receive point, possibly their transmitter tower or another high point in town, from where the signal is captured from the remote site and relayed to the studios.

At the remote site, such equipment is usually mounted on an electronic news gathering (ENG) truck, as shown in Figure 11.3. The truck also provides transport for the camera crew and equipment and usually has an electrical generator so it is self-contained.

Upon arrival at the news site, the link equipment operator raises the antenna mast and aims the microwave transmitting antenna in the direction of the receive antenna. For a live story, the video output of the camera and the sound mixer audio (or reporter's wireless microphone receiver output) are fed to the microwave transmitter. If the story is prerecorded, either the camcorder playback output is fed to the link, or the camera recording may be replayed from a player in the truck.

It is also possible to use a microwave link with a helicopter being used for live ENG, in which case a stabilized, steerable antenna is required to maintain the line-of-sight link to the ground receive point.

FIG. 11.3.   ENG Microwave Truck. Courtesy of Frontline Communications

Microwave links can carry both analog and digital audio and video signals. ENG links for DTV are available using COFDM modulation (see Chapter 4) to provide a very robust signal. Although such links must be licensed to broadcasters, this does not in itself ensure interference-free operation, since several broadcasters may be licensed to use the same frequency in a given area. Therefore, there must be good coordination between all operators of ENG microwave equipment in a given geographical area so their microwave links do not interfere with one other. This process is called *frequency coordination*, and in the U.S. it is often provided by local members of the Society of Broadcast Engineers (SBE). It becomes particularly crucial when a location becomes the center of a national event or news story, and multiple broadcasters from the local area and elsewhere converge on a single site, each seeking spectrum for their contribution links back to their studios or production centers.

Satellite

Microwave links to the station do not work for all locations where news may occur. To allow live reporting over a larger area, TV networks and some stations use *satellite news gathering* (SNG). This refers to the satellite link used for sending the signal to the studio, which is used in lieu of the terrestrial microwave paths described above. An example of an SNG truck, with its *uplink* dish antenna, is shown in Figure 11.4.

The signal from the satellite is received at the station (or wherever else it is needed) with a satellite downlink. Various frequencies in the SHF band

FIG. 11.4. SNG Uplink Truck. Courtesy of Frontline Communications

are allocated for these links, with separate frequencies used for *Earth-to-space* (uplink) and *space-to-Earth* (downlink) channels. Different bands are denoted by alphabetic letter combinations, with the most commonly used by broadcasters being *C-band*, *Ku band* and *Ka band*. Because C-band uses lower frequencies—and therefore longer wavelengths, as described in Chapter 7—its dishes are larger (approximately 10 feet in diameter), whereas the K-band dishes are typically smaller (approximately 2 to 5 feet in diameter).

Video-over-IP

A more recent development takes advantage of the proliferation of IP network access, via both wired and wireless means. While not necessarily as reliable as the links described above, the low cost and high convenience offered by *Video-over-IP* has made it quite popular for breaking news coverage, for example, where timely reporting is of the essence, and lead time is short to non-existent.

In such cases, the remote reporter or crew feeds audio and video into an interface device that includes audio and video codecs along with interfaces to wired or wireless IP links. In the latter case, the links may be provided by one or more existing broadband Internet smartphones or cellular data modems (or "dongles"), operating on so-called *3G* or *4G* wireless networks (WANs), which are connected to the Video-over-IP interface device via USB or a similar connection. Figure 11.5 shows an example.

FIG. 11.5. Video-over-IP Interface Mounted on a Camera, and Connected to Two 4G Wireless Data Modems on USB Adapters. Courtesy of Comrex Corporation

NETWORK DISTRIBUTION LINKS FOR RADIO AND TELEVISION

Satellite

The majority of radio and television networks and syndication program services distribute their content to affiliates via geostationary satellites. Signals are received at the station with Earth-station downlink antennas of a type similar to that shown in Figure 11.6. As with SNG, various frequencies in the SHF band are allocated for this purpose.

The networks and syndicators do not own or operate these satellites, but rely on third-party communication companies to deliver this content.

LAN/WAN

Network programming that is not for live broadcast may be received as audio and video files distributed over an IP-based computer network, just as described above for contribution links. In some cases, even live content may be distributed in this fashion, via robust IP streaming techniques, generally via dedicated bandwidth on terrestrial or satellite IP paths.

Most network distribution systems also include a communication link, to allow network centers to alert affiliates of upcoming transmissions or last-minute

FIG. 11.6.   Satellite Downlink Antenna. Courtesy of
Patriot Antenna Systems

changes. One important element of such communications is notification of where insertion points will be for stations to run local content within a network program.

STUDIO-TRANSMITTER LINKS FOR RADIO AND TELEVISION

The STL is the means by which program material produced in a local station's studio facility is transported to the station's transmitter site(s) for broadcast. If the station's transmitter is located adjacent to its studios, then the STL might simply be a cable, or set of cables, connecting the output from master control to the transmitter area. More often, however, the transmitter is placed at a remote location, such as the top of a mountain or tall building, or on a tower across town. In those cases, there are several different methods for transporting the program signal, as described in the following sections.

Leased Telephone Lines

One traditional means of getting program signals to the transmitter is to use leased telephone-company provided lines. The local telephone company may be able to provide broadcast-quality links between studios and transmitters so that the broadcaster can feed the program audio and video signals—along with control signals for the transmitter and other equipment located at the transmitter site—into telephone-company interface hardware at the studio facility, and retrieve them from a telephone company terminal at the transmitter facility. A range of technologies may be used by the telephone company to provide this service to radio and television stations, but at this writing these services all use some form of digital transmission, and increasingly via IP. These links are typically bidirectional (allowing the station to monitor transmitter equipment and return contributed content that may be picked up from an RPU receive antenna located at the same tower site, for example), although they are often *asymmetrical*, meaning that the bandwidth from the studio to the transmitter is higher than the bandwidth provided in the opposite direction (the latter is sometimes called *TSL*, for *Transmitter-to-Studio Link*).

ISDN Line

ISDN lines, as discussed under Contribution Links, can also be used as an STL (typically only for radio), although because the service is often billed on a per-minute of usage basis, they are often used only as a backup to a primary STL.

T1 and Similar Lines

A T1 line (also known as a *DS1* line, telephone-company nomenclature for "Digital Signal 1"), as discussed under Contribution Links, is suitable as a radio

STL, assuming that it is possible to get the wideband connection to the remote transmitter site. It may be leased from a telecommunications provider as a wired service, or implemented as part of a station's own (licensed) broadband microwave system.

Similar, scaled-up services at higher bandwidths can be used in the same ways for television STLs. For example, DS3 service—also known as a T3 line—operates at 45 Mbps, and can be used for this purpose to send multiple signals in either direction between the studio and transmitter sites. (A DS3 line is composed of 28 DS1 lines bundled together.)

Microwave Links

Terrestrial microwave STLs may be used for both radio and television, based on the same principle as television ENG contribution links. In this case, they will have fixed microwave transmitters and antennas at each end of the link, operating under licensed authority. Because microwave transmissions work only with direct line-of-sight, the transmitting antenna must have a good "view" of the receiving antenna at the transmitter site. If an obstruction is unavoidable, sometimes a *multihop* link is installed, with a microwave receiver and second transmitter at an intermediate location that can see both the studios and transmitter sites. Longer or more complex multihop links may have more than two segments, and in some cases, may drop the signal off to additional broadcast transmitters along the way—for example, in a state-wide or regional distribution network to a group of stations all running common content.

Examples of typical microwave antennas on a tower are shown in Figure 11.7. With fixed links, the actual radio transmitter and receiver equipment is often located in a building adjacent to the tower, connected to the antenna with an RF feed-line. This type of link is widely used for both radio and television. They are available with the wide bandwidth needed for video signals, and are good for linking to the remote mountaintop locations where FM and television transmitters are frequently located.

Fiber-Optic Links

The actual physical path used by a telecommunications service provider for the fixed services described above may be optical fiber rather than copper twisted-pair or coaxial wiring. In some locations, it also may be possible for a broadcaster to run its own fiber-optic path from the studio facility to the transmitter site, or to lease so-called "dark fiber" from a third party, which the broadcaster then "lights up" with its own terminal equipment.

Optical fiber can provide excellent, high-quality, bidirectional links for both audio and video, over long-distance paths. A special codec at each end of the fiber combines the various signals to be sent on a single fiber.

FIG. 11.7. Microwave Link Antennas. Courtesy of Microwave
Radio Communications

Fiber-optic cable consists of a long, thin strand of glass, sheathed in plastic, possibly with multiple fibers in each sheath. A beam of special light from a laser is modulated with the digital signals to be carried in a multiplexed arrangement. The light enters the fiber at one end and undergoes repeated *total internal reflection*, which allows it to propagate for very long distances with negligible loss and no distortion until the optical receiver at the other end ultimately detects it. Another advantage of fiber-optic cable is that the light signals are largely unaffected by electrical and magnetic interference.

There are two main varieties of fiber-optic cable: single-mode and multimode, which relate to the way the light is reflected inside the strand of glass. Single-mode is generally more expensive, but is required for longer distances.

TRANSMISSION STANDARDS
AND SYSTEMS

CHAPTER 12

Analog Radio

In the United States, the Federal Communications Commission (FCC) regulates all radio broadcast transmissions. The FCC stipulates what standards or systems must be used and licenses stations to operate on particular frequency allocations at specific power levels. Stations within a certain distance of the borders with Canada and Mexico must also comply with related treaties and agreements made by the United States with the governments of those countries.

There is no named standard for basic AM and FM analog radio transmissions. The modulation and transmission systems use principles and technologies that are well defined and used throughout the world, with some minor variations. The FCC rules, however, specify various parameters, such as the spectrum and channel bandwidths to be used in the U.S. for AM and FM radio broadcasting. These rules enable U.S. stations to coexist with one other and allow receivers to operate in an optimum fashion. In addition, the National Radio Systems Committee (NRSC) has developed several standards that are referenced in the FCC rules, which further define and specify the characteristics of today's AM and FM broadcast signals.

As discussed in Chapter 7, information is carried on a radio frequency *carrier wave* by modulating one or more of its attributes. For the different types of *modulation* used in radio broadcasting, this chapter looks at *carrier frequencies* and *sidebands*, their relationship to broadcast radio *channels*, and the use of *subcarriers* to carry additional information.

AM TRANSMISSION

In transmissions using AM—*amplitude modulation*—the program audio signal is used to modulate (vary) the amplitude of the carrier wave that will be transmitted by the station. When the amplitude of the program signal is zero, the carrier remains unmodulated. As the instantaneous amplitude of the program signal increases up to its maximum, then the carrier amplitude varies accordingly, up to the maximum amount possible under FCC rules, which corresponds to 100 percent modulation.

AM services are very susceptible to *interference* from outside sources of RF, such as car ignitions, electric motors, fluorescent lights, or other devices. This is because the interfering signals add or subtract from the amplitude of the wanted RF signal that carries the audio from the radio station, and the AM radio receiver cannot distinguish between the radio station's transmission and the interference sources.

The limited bandwidth allocated to AM channels constrains the bandwidth of the audio signal to about 10 kHz. Most AM receivers actually have much lower high-frequency response than this, sometimes 5 kHz or less, depending on various factors. Although this restricted bandwidth is fine for speech, music tends to sound muffled and lacking in treble. Combined with the noise and interference that may occur on AM, this means that stations wishing to broadcast high-quality music generally prefer to use FM, which does not suffer from these problems (as explained below). On the other hand, AM signals have the advantage that they may often be received more easily than FM over a large area, particularly at night (see Chapter 20).

Carriers and Channels for AM

AM broadcasting in the U.S. uses the frequency band of 535 to 1705 kHz, which is considered the medium frequency (MF) band. The carrier wave transmitted by each AM station's transmitter is assigned (by an FCC-issued license) to a specific channel, which is identified by its center frequency. AM stations in the U.S. are spaced at 10 kHz intervals (other countries use either 9 kHz or 10 kHz spacing). Thus, the lowest U.S. AM radio channel frequency available is 540 kHz, which nominally occupies the bandwidth of 535 to 545 kHz.

The next channel up the band is 550 kHz, occupying 545 to 555 kHz, and so on. Note that for protection from interference, immediately adjacent channels are generally not assigned to the same geographic area (in any broadcast band). In some cases, two or more channels are left vacant between channels assigned to operate in the same geographic area. The terminology used for two stations immediately next to each other in any broadcast band is *first adjacent*, while two stations that have one open channel between them are called *second adjacent*, and so on. Stations occupying the same frequency (in different geographic locations) are called *co-channel*.

This is a particular issue for AM, because audio bandwidth is allowed to extend to 10 kHz, meaning that actual occupied bandwidth of a station is 20 kHz (i.e., 10 kHz above and 10 kHz below the assigned center frequency), which extends halfway into the station's first-adjacent channels. FCC allocations of AM channels take this into account, however, and recent industry recommendations and practice have attempted to voluntarily limit AM audio bandwidth to 5 kHz in an attempt to further reduce interference. Not all AM broadcasters comply with this recommendation, however. (More on this can be found in Chapter 13, under AM IBOC.)

Sidebands and Bandwidth

In AM broadcasting, when the modulating signal (the audio program) is combined with the carrier, the varying-amplitude RF signal that results is made up of the original carrier frequency and additional frequencies known as the upper and lower *sidebands* (see Chapter 7). As described above, if the audio signal has frequencies extending to 10 kHz, then the sidebands will extend about 10 kHz above and below the carrier frequency, occupying at least 20 kHz of radio spectrum. In reality, transmitters may also produce additional energy in sidebands extending beyond these limits, which is referred to as *splatter*.

The maximum RF emission bandwidth allowed is specified by the FCC (see below), but almost inevitably, both the first- and second-adjacent channels to any AM station cannot be used in the same geographic area as that station. Other factors affecting interference also include a station's transmitted power and the directional characteristics of its transmit-antenna system.

EMISSIONS MASKS

Figure 12.1 shows the AM and FM band *emissions masks* defined by the FCC to protect other stations operating on nearby channels from interference. The emissions mask is the limit placed on the signal strength of the broadcast signal, and it is defined over a range of frequencies surrounding the carrier frequency. A broadcast station's signal strength at specific frequencies must decrease as the frequencies become farther away from the carrier. FCC masks define this attenuation vs. frequency in a step-wise fashion, with the first step down shown by the −25dB (decibel) value shown in the figure.

As Figure 12.1 also illustrates, the radio spectrum available to an AM station (about 20 kHz) is approximately 10 percent of that available to an FM station (about 240 kHz).

Subsidiary Communications for AM

An AM station's program material occupies all or most of the radio spectrum assigned to the station. This makes it extremely difficult to place a subcarrier for additional services in the channel, as used for FM stations (see following section).

Therefore, another method may be used for *multiplexing* (i.e., combining together) two signals in an AM broadcast channel. *Quadrature amplitude*

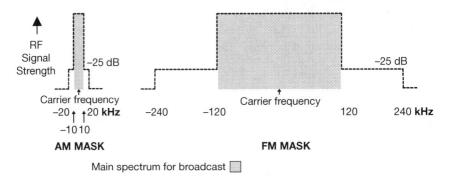

FIG. 12.1. Basic View of AM and FM Broadcast Channel Emissions Masks

modulation (QAM) has been used for many years as a means of allowing AM broadcasters to transmit auxiliary information on the main carrier frequency by varying another parameter of the carrier signal called *phase* (see Chapter 7 for more on QAM). For a time, this approach was used to allow AM stations to broadcast in stereo, but the format did not succeed in the market, so it has largely been abandoned. (Subsequently, IBOC digital radio has allowed AM stations to offer stereo audio – see Chapter 13.)

FM TRANSMISSION

In transmissions using FM—*frequency modulation*—the program audio signal is used to modulate the <u>frequency</u> (as opposed to the amplitude) of the carrier wave that will be transmitted by the station. When the amplitude of the audio signal is zero, the carrier remains unmodulated. As the instantaneous amplitude of the audio signal increases up to its maximum, then the carrier frequency varies accordingly, up to the maximum amount allowed by regulation, equivalent to 100 percent modulation.

FM services are very robust and immune to interference from outside sources of RF. This is because, although interfering signals may add or subtract from the <u>amplitude</u> of the RF carrier (as they do with any RF signal), they do not affect the <u>frequency</u> of the wanted signal that carries the audio information heard by an FM receiver.

As mentioned previously, the bandwidth allocated to FM channels is also much wider than AM, and this allows the bandwidth of the audio signal that can be transmitted to extend to about 15 kHz. This wide frequency response, combined with high *signal-to-noise* ratio and low interference, makes FM capable of high-quality audio transmission.

Carriers and Channels for FM

For FM broadcasting, the range of RF frequencies assigned in North America (and in some other regions) is 88 to 108 MHz, which is in the Very High Frequency (VHF) band. FM broadcast carriers are assigned to channels, which are spaced at 200 kHz intervals in the United States (other countries use either 100 kHz or 200 kHz spacing), with the carrier frequency again in the center of each channel.

Deviation, Sidebands, and Bandwidth

The change in frequency of the FM carrier as it is modulated is known as the *deviation*, or *frequency swing*. The maximum deviation allowed for 100 percent modulation in the U.S. is specified by the FCC as plus and minus 75 kHz.

Because it is the frequency, and not the amplitude, of the carrier that is varied, FM produces sidebands in a different way from AM. As a result, the modulating signal of an FM carrier produces a range of upper and lower sidebands, extending much more than plus or minus 75 kHz from the carrier frequency. When using subcarriers (see below), the sidebands extend out further still. As Figure 12.1 illustrates, the radio spectrum available for the main signal from an FM station is specified by the FCC as plus or minus 120 kHz from the carrier frequency.

Because FM channels are allocated at 200 kHz spacing, the nominal 240 kHz bandwidth of the transmitted signal from an FM station therefore extends into adjacent channels both above and below its allocated frequency (although not to the same extent as with AM broadcast channels). As with AM broadcasting, the maximum RF emission bandwidth allowed is specified by the FCC, and is affected by a station's transmission power and the directional attributes of its transmit antenna (although directional transmission antennas are far less common among FM stations than they are for AM stations). To mitigate interference, however, stations are generally not assigned to first adjacent FM channels in the same geographic area.

STEREO CODING

Two-channel stereo sound, consisting of left and right program channels, is used almost universally at FM radio stations. As noted above, a system of transmitting stereo over AM was developed and incorporated into the FCC's rules, but because of increased potential for interference with other stations and the generally poorer audio performance of AM, the system was not widely adopted.

The approach used to broadcast analog stereo FM ensures that the stereo signal can be decoded and played by both stereophonic and monaural receivers. A similar approach was used by AM Stereo, and also in the Multichannel Television Sound system used in the latter days of NTSC analog television broadcasting, where the audio format was called the *BTSC System*, after the name of group that created the standard, the Broadcast Television Systems Committee. (BTSC also included an additional *Second Audio Program*, or SAP, channel—typically used for an alternate language soundtrack—and a communications channel for broadcasters called the PRO channel.)

In all of these cases, monaural audio had preceded the introduction of stereo, so for maximum efficiency, the newer, stereo audio format had to made *backward compatible* to mono, so legacy receivers that were capable of only monaural audio would be properly served, and new receivers with stereo capability could obtain stereo audio. For such stereo/mono compatibility, it is not sufficient to simply transmit separate left and right channel signals on the broadcast carrier. Instead, a "main program" channel must be transmitted that combines both the left and right audio signals together, placing it on the carrier in the same place as regular (mono) audio had traditionally been, so it can be used by a monophonic receiver without adaptation. Meanwhile, a new, "stereo program" channel must be transmitted that can be coupled with the main program channel in new stereo receivers to extract separate left and right audio program material.

Figure 12.2 illustrates the method used to achieve this result. Before broadcast transmission, the left and right audio channels are added together to produce a combined mono *sum* signal (left plus right), and the right channel signal is subtracted from the left to produce the supplementary stereo *difference* signal (left minus right). (The process of "subtracting" one signal from another electronically is simply performed by inverting the polarity—or exchanging the "plus" and "minus" wires—of the subtracted signal, and adding it to the other signal.) After

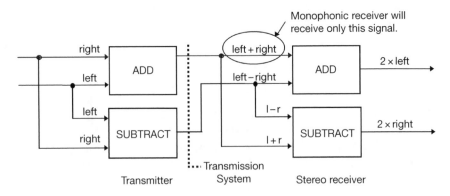

FIG. 12.2. Stereo Multiplex Coding

passing through the transmission system, the sum ("left plus right," or main audio program signal) can be received by mono receivers and played over a single loudspeaker. At a stereo receiver, the sum and the difference ("left minus right" or stereo subchannel) signals are themselves added and subtracted from each other, as shown in Table 12.1 below. This process is sometimes called *matrixing*. The left column of the table shows the result of the receiver's adding the sum and the difference channels to each other, and the right column shows the result of the receiver's subtracting the difference channel from the sum channel. Note that the difference signal $(L - R)$ when inverted for subtraction from the sum signal has its signs reversed to become $(-L + R)$.

This produces the original individual left and right channel signals (actually at twice the amplitude, as indicated by 2L and 2R, but that is easily adjusted), which can be played over stereo loudspeakers or headphones.

The stereo coding process is carried out at a broadcast station using a *stereo generator*. Besides applying the sum and difference process to the left and right audio channels, the stereo generator also places the resulting audio signals onto the broadcast carrier in a special way, as described next. For FM radio (the only remaining broadcast service where this process is still in frequent use), the system is sometimes called the *Zenith system* because it was developed by the Zenith Radio Corporation, or more generally, the *FM stereo multiplex system*.

Stereo Signal Generation

The stereo generator at an FM station produces a *composite* output signal as shown in Figure 12.3. The "left plus right" main program signal needed for mono reception is baseband FM audio, with 15 kHz bandwidth. A *stereo pilot* tone at 19 kHz is added as a reference and synchronization signal (to keep the sum and difference signals in time alignment), and the "left minus right" stereo difference signal needed to produce the stereo channels is <u>amplitude modulated</u> onto the 38kHz *second harmonic* of the 19 kHz stereo pilot (i.e., a subcarrier at twice the

TABLE 12.1

	Receiver Matrix SUM	Receiver Matrix DIFFERENCE
Main Audio Program	L + R	L + R
Stereo Subchannel	L – R	L – R
Result	L + R + L – R = *2L*	L + R – L + R = *2R*

Table. 12.1.   Stereo Multiplex Matrix.

frequency). Because the stereo difference signal is (like the main program) an audio signal with 15 kHz bandwidth, the modulated subcarrier has lower and upper sidebands, carrying the program information, which extend from 15 kHz below to 15 kHz above the 38 kHz subcarrier, as shown in Figure 12.3. This whole composite stereo signal, which is still at baseband, with a 53 kHz bandwidth, is fed to the FM transmitter.

Although now all combined ("multiplexed") on a single carrier, these different signals do not interfere with each other because they each occupy different frequencies on the carrier. This is how the stereo receiver can easily separate the signals out again for decoding. A mono radio receiver will not be able to separate out the stereo subcarrier signal from the main baseband audio (mono) signal, but this is of no consequence because the stereo subcarrier is not needed by the mono receiver, and it is above the top of the range of human hearing so it is not reproduced by the receiver.

The stereo pilot mentioned earlier is an unmodulated subcarrier (i.e., a subcarrier that does not have any additional information added to it), so it is a pure, sine-wave tone, with no sidebands, therefore occupying no additional bandwidth other than its own frequency of 19 kHz. As mentioned above, one purpose of the stereo pilot is to provide a reference frequency for the receiver to demodulate the stereo subcarrier. It also tells receivers that the host FM station is indeed broadcasting in stereo. If an FM station does not transmit the stereo pilot signal, then stereo receivers will assume that the station is broadcasting a monaural program, and they will not try to decode the received audio into separate left and right stereo channels. Stereo receivers usually have some visual indicator on their control panel to show the detection of a stereo pilot.

Because the stereo subcarrier uses AM to carry the stereo difference information as described above, whereas the main (mono) audio channel is at baseband FM, there is additional noise introduced in the received stereo signal compared to the mono signal. When the FM station's received signal is weak, this noise

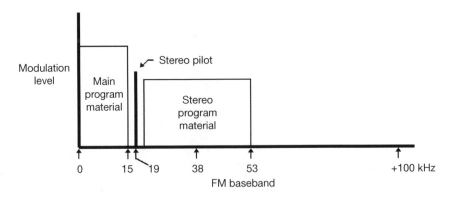

FIG. 12.3. Composite FM Stereo Signal

differential can become worse, so many stereo FM receivers will only decode the stereo audio when the RF signal is relatively strong, to avoid introducing excessive noise in the received stereo signal. In such cases, the stereo receiver automatically defaults to the main audio only, allowing continued listening at reasonable quality in mono. This feature is called *"blend,"* and when activated, the stereo indicator on the receiver is usually turned off.

SUBCARRIERS

The subcarrier approach used in FM stereo multiplexing can also be used to carry other information in a broadcast signal, which may or may not be associated with the broadcaster's main programming. The following discussion about subcarriers uses the FM baseband signal for illustrative purposes, but the theory could be applied to any RF carrier.

A *subcarrier* is a special type of RF carrier. Like any carrier, it uses some form of modulation to carry information, but the term "subcarrier" implies that it is one of several that are combined within a given channel, or that it is added to a host carrier before the combined signal is modulated in the transmitter in order to be delivered to a receiver (as described in Chapter 7). Subcarriers are common in FM radio—as they were in analog TV systems—because these channels were designed to have extra bandwidth for the future addition of new features, and subcarriers were a common method of doing so. (If properly designed, subcarriers can be added to a system without disrupting legacy receivers, since those receivers will simply ignore them, and new receivers can be designed to detect them within the existing channels. This is how stereo audio was added to the FM radio broadcast system, which was originally a monaural format.)

Figure 12.3 illustrates the fact that an FM stereo audio signal actually occupies only a little more than half of the baseband spectrum that can be used. The excess channel capacity (from 53 kHz where the stereo subcarrier's bandwidth ends to the upper limit of 100 kHz on the FM channel's baseband) that remains presents an opportunity for FM broadcasters to offer additional services.

In addition to the 38 kHz stereo subcarrier, there are three other FM broadcast subcarriers in use today. They are the 57 kHz subcarrier used for the *Radio Data System* (RDS, see below), a 67 kHz subcarrier, and a 92 kHz subcarrier, as shown in Figure 12.4.

The technical specifications of the 67 and 92 kHz subcarriers are regulated by the FCC, but their applications are not, so broadcasters have used these for a variety of services, both commercial and non-commercial. The FCC originally referred to these as *Subsidiary Communications Authorization* (SCA) services, and for many years FM broadcasters have generated additional income by leasing out these subcarriers to third parties for services such as background music distribution, radio paging, or private datacasting. Another application is "narrowcasting," which involves distribution of services to special interest groups, such as radio for ethnic communities or Radio Reading Services for visually impaired listeners, in which newspapers and other printed material are read by (often volunteer) announcers. In all such cases, special subcarrier-capable receivers are used by the narrowcast communities, which are tuned to the station's carrier frequency, but which demodulate and decode the 67 or 92 kHz subcarrier rather than the station's main program from the FM channel.

Radio Data System (RDS)

In 1993, the National Radio Systems Committee (NRSC) adopted a standard for transmitting digital data at 1187.5 bits per second on a subcarrier in the FM baseband. This standard is called the *United States Radio Broadcast Data System* (RBDS), although it is often referred to simply as RDS, for *Radio Data System*. RDS was originally a European standard, and it was adapted for U.S. use by the NRSC as RBDS. Nevertheless, consumer electronics manufacturers refer to both systems generically as "RDS." Updated editions of the RBDS standard have been subsequently adopted by the NRSC in the interim.

The RDS signal can carry up to about a dozen different types of data related to the station and its content, such as station identification, station format, song titles, artist names, and other program- or station-related information. It can also accommodate other text such as weather and traffic updates, or emergency alerting. RDS data can be displayed as static or scrolling text on a receiver's display panel. It has allowed analog FM broadcasters to add *metadata* to describe their services and/or content, and thereby compete with digital services that inherently include such additional descriptive material.

The RDS subcarrier is centered on 57 kHz, the third harmonic of 19 kHz (the FM stereo pilot frequency), and the recommended bandwidth is approximately 4 kHz. As shown in Figure 12.4, the RDS signal fits between the stereo program material and the 67 kHz subcarrier.

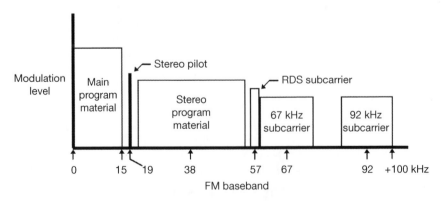

FIG. 12.4.   FM Stereo Signal with 57 kHz (RDS), 67 kHz, and 92 kHz Subcarriers

CHAPTER 13

IBOC Digital Radio

In-Band, On-Channel (IBOC) is a digital radio technology deployed in the United States and elsewhere, whereby a digital signal is combined with the signal from an analog station. The name IBOC indicates that the digital service shares the same band and channel as the analog service. The technology is more commonly known today by its commercial name, *HD Radio*, which is a trademark of the system's developer, *iBiquity Digital Corporation*.

Both an AM ("AM-IBOC") and an FM ("FM-IBOC") system are in use. They provide similar capabilities, although the technical details vary. Characteristics that are common to both systems are described as follows.

PHASED IBOC INTRODUCTION

IBOC is designed to be introduced in two phases—*hybrid* and *all-digital*—enabling a smooth evolution from the current analog radio broadcasting.

Hybrid Phase

The initial *hybrid* phase adds a digital service to the existing analog service, with identical audio programming (i.e., simulcasting of the analog service), additional metadata, and some optional services (which may include additional audio channels). This provides simultaneous analog and digital services to approximately the same principal coverage area. Because the digital portion of the IBOC signal is at a much lower power level than the analog portion, as listeners approach the edge of the coverage area, the digital signal may be lost prior to the loss of the analog signal. If that occurs, the IBOC receiver *blends* from the digital to the analog signal. This avoids the abrupt reception failure that would otherwise occur (the so-called *cliff effect*), and maintains the main (analog) program's continuity, although metadata and optional services unique to the digital signal (more on these below) are lost after blending.

During the hybrid phase, listeners may continue to use their existing receivers for the analog audio service or may purchase digital receivers for the digital audio and new services. The digital units also remain capable of receiving stations that are only transmitting analog services without IBOC signals.

All-Digital Phase

At some point in the future, when IBOC receiver penetration is sufficient, it is envisioned that broadcasters would be in a position to implement the *all-digital* phase. At that time, the analog signal could be turned off and the digital signal enlarged to fill the channel capacity now no longer used by the analog signal (but with all emissions remaining within the FCC mask). This maximizes the digital coverage area and will provide capacity for additional services from each broadcaster, while remaining on the originally assigned broadcast frequency. Currently, however, the FCC does not allow broadcasters to utilize this mode of operation.

Note that if a station were to use an all-digital IBOC transmission mode, legacy analog-only radios would no longer receive the station's signals. Therefore as of this writing, it seems unlikely that broadcasters would choose to pursue this course anytime soon. Nevertheless, for AM radio, the interference and ever-increasing noise that plagues reception of most stations makes an all-digital AM-IBOC approach seem less farfetched. At this writing, this is one prospect being considered for *AM revitalization*, a process exploring various solutions to ameliorating or eliminating the current technical difficulties of AM radio reception. (See Chapter 20 for more on this issue.)

CARRIERS AND CHANNELS FOR IBOC

For hybrid IBOC, the digital service is achieved by placing additional carriers in the sideband areas around the analog signal of the associated AM or FM station (see Figures 13.1 and 13.2 below). These carriers are arranged in a special way so as to minimize the interference to the analog portion of the signal. The digital signals fall within the existing emission masks of the AM or FM channel, as described in the previous chapter. In the all-digital mode, the analog audio service is removed and replaced with additional digital carriers in the center of the channel.

MODULATION AND FORWARD ERROR CORRECTION

IBOC uses a multi-carrier modulation technique called *coded orthogonal frequency division multiplexing* (COFDM) to carry the digital signal (see the Modulation section in Chapter 4 for a brief description of COFDM). It is the same system used by other telecommunications systems, including most wireless phone systems. Its main advantage for IBOC broadcasting is that, when used in conjunction with techniques such as *interleaving* and *forward error correction* (FEC), COFDM makes the signal very robust and easier to receive under difficult conditions, particularly in a moving vehicle.

AUDIO DATA COMPRESSION

Both AM and FM IBOC rely on lossy data compression to reduce the data rate of the digital audio to a rate that can be carried reliably in the available broadcast bandwidth. The audio coding system ("codec") used for IBOC is known as HDC.

As described in Chapter 5, typical uncompressed stereo audio at the IBOC standard sampling frequency of 44.1 kHz with 16 bit resolution (as used for CDs) has a data rate of about 1.4 Mbps. This has to be reduced to no more than 96 kbps for FM-IBOC and 36 kbps for AM-IBOC. Like any codec, HDC performs this bit-rate reduction by allowing certain parts of the audio signal to be discarded, or sent with less accuracy, without significantly changing the sound heard by the listener (as detailed in Chapter 5).

AM IBOC

Sidebands and Bandwidth

The RF channel arrangement for AM hybrid IBOC is shown in Figure 13.1. The analog audio sidebands, carrying the mono analog program, extend to about 5kHz on each side of the carrier, and the *digital sidebands*, carrying the digital service, extend out to just under 15kHz. Although the term "digital sideband" is frequently used, they are not really sidebands in the traditional sense; they are actually just the areas of spectrum above and below the analog host where separate digital carriers are placed. The primary, secondary, and tertiary "sidebands" carry different parts of the digital signal. As shown by the FCC AM Mask line, the digital carriers are at a much lower level than the analog signal, which helps reduce levels of interference into the analog host signal as well as into adjacent AM channels (see also Figure 12.1 in Chapter 12).

In AM-IBOC, the digital carriers are modulated using QAM (see Chapter 7). This approach allows these carriers to also overlap the part of the channel used for the analog program service, as shown in Figure 13.1.

Nighttime Operations

The AM radio band in the United States is congested with many stations. Because of the way signals in the MF band used by AM broadcasting are propagated over

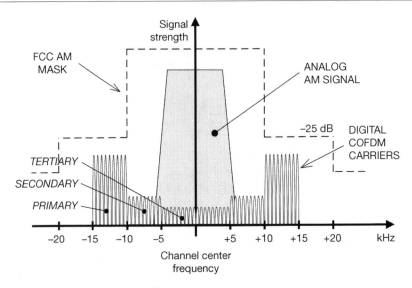

FIG. 13.1. Hybrid AM-IBOC RF Channel

long distances after sunset, nighttime operation for AM stations is limited by interference from, and to, quite distant stations. This extended nighttime propagation is called *skywave*. At the time of the FCC's interim authorization of IBOC in 2002, concerns about how AM-IBOC could increase skywave interference caused the Commission to allow AM-IBOC operation only during daytime. In 2007, however, the FCC changed its rules to allow full-time AM-IBOC operations. (See Chapter 20 for more information on skywave propagation.)

Quality and Bit Rates

Due to the presence of the digital carriers, when using IBOC the AM analog audio signal bandwidth has to be limited to approximately 5 kHz. Although as noted in Chapter 12, analog AM stations can theoretically carry audio up to about 10 kHz, in fact most current AM receivers cut off at frequencies below 5 kHz, so most listeners are not likely to notice this reduction in analog bandwidth. AM-IBOC is not compatible with the analog AM stereo format; therefore, any AM stations still broadcasting in stereo will need to revert to mono for their analog service if they intend to broadcast IBOC.

The AM-IBOC digital signal provides stereo audio at a quality comparable to existing FM radio, using audio compressed at 36kbps. As a listener gets farther away from the IBOC transmitter, the AM IBOC digital audio signal blends from 36kbps stereo to a 20kbps mono audio that is carried as a more robust signal.

FM IBOC

Sidebands and Bandwidth

The RF channel arrangement for FM hybrid IBOC is shown in Figure 13.2. The analog audio sidebands extend to just under 130kHz on each side of the carrier, and the "digital sidebands" extend out to just under 200kHz. Conforming to the FCC FM mask line, the digital carriers are at a much lower level than the analog signal, which helps minimize interference into the analog "host" as well as the adjacent channels.

The original (2002) FCC rules regarding FM-IBOC broadcasting set the power levels for the digital carriers quite conservatively, at no higher than 20 dB below the analog carrier's power (i.e., running the digital signal at about 1/100 of the analog power). However, in 2010, the FCC amended these rules to allow stations to increase digital power by up to 6 dB, thereby quadrupling it (to 14 dB below analog). It also allowed stations to apply for even higher digital power authorizations of up to 10 dB below analog (i.e., 1/10 the analog power), which the FCC would review and authorize on a case by case basis, if it were satisfied that the increase would not cause unacceptable adjacent-channel interference. This allowed the digital service areas of FM-IBOC stations to more closely match their analog service areas.

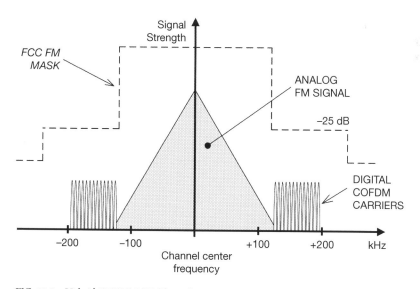

FIG. 13.2. Hybrid FM IBOC RF Channel

Quality and Bit Rates

Unlike AM-IBOC, no new restrictions are placed on the FM analog stereo signal when using IBOC. The FM IBOC digital *main program service* (MPS) can provide stereo audio at near-CD quality, using compressed audio at a transmitted data rate of up to 96 kbps. The FCC does require the quality of the IBOC MPS to be at least a high or higher than that of the analog FM service.

There is also an optional mode of operation for FM IBOC called *extended hybrid* in which the bandwidth used for digital carriers is enlarged (toward the channel's center frequency, overlapping a bit into the analog signal area), allowing the IBOC signal to provide an additional 24 kbps of payload, thereby allowing support for more audio or data services.

Supplemental Program Services

It is possible to subdivide the 96 kbps IBOC-FM hybrid payload into two or more audio streams, say one running at 64 kbps and another at 32 kbps, to provide a *supplemental program service* (SPS). The quality for the two channels can still be very good, and this gives stations the opportunity to broadcast different programs simultaneously. Some broadcasters have subdivided further to include additional supplemental channels. The system can support up to four digital audio program services. Unlike the main program service, however, SPS is digital-only, with no analog backup for blend in the hybrid FM-IBOC mode.

These additional services are commonly referred to as *multicasts.* (In some contexts they are specified as *HD Radio multicasts,* to distinguish them from the analogous DTV multicasts enabled by the ATSC system, or from the quite different *IP multicast* format used on the Internet and in AoIP or Video-over-IP systems, as described in Chapters 6, 8 and 9, respectively.) A more common nomenclature for such services is *HD-1, HD-2, HD-3* and *HD-4,* which is how they are typically identified on FM-IBOC receivers. In this case, "HD-1" refers to the MPS, which must be the digital simulcast of the FM station's analog signal (as required by FCC rules, if the station chooses to operate digitally), while "HD-2" refers to the first multicast service (SPS1), "HD-3" refers to the second multicast service (SPS2), and so on.

PROGRAM AND SERVICE DATA (PSD)

In addition to audio content, FM-IBOC provides metadata to receivers that describe elements of the station's service and its currently airing content. Called *Program and Service Data* (PSD), this feature allows the station to display textual information about itself on IBOC receivers, as well as to provide substantial text data describing the audio content it is broadcasting (separately on each multicast channel, if used), such as song title, artist and album. Unlike the RDS system described in Chapter 12, IBOC PSD provides far less constrained data capacity, and much faster display of text on the receiver. Many FM-IBOC broadcasters feed the same or similar metadata to both RDS and IBOC-PSD encoders, so both analog and digital receivers can benefit from the metadata display. Most FM-IBOC receivers default from PSD to RDS display when they blend from MPS digital to analog audio, as well.

DIGITAL RADIO DATA BROADCASTING

Data Services Using Hybrid IBOC

Besides PSD, FM-IBOC stations can also accommodate transmission of other data not specifically related to the station or current audio content, such as weather and traffic updates. In some cases, carriage of this data for third parties allows stations to earn additional revenue. Like multicasting, however, the bandwidth allocated to such additional data reduces the payload bit rate that remains available for regular station content.

Data Services Using All-Digital IBOC

The all-digital mode of IBOC provides more space for new data services. This is particularly valuable for AM-IBOC, in that it would allow PSD and third-party datacasting, which is not supported in the AM-IBOC hybrid mode.

ADVANCED SERVICES

New features continue to evolve for the IBOC system. At this writing, a recently developed capacity allows graphical metadata to be broadcast, which provides a station the capacity to move beyond simple text display to include line art (e.g., station or advertiser logos) and images (e.g., album-cover art or artist/DJ photos). The latter application has given rise to this feature being called *Artist Experience*. It is expected that additional features will continue to be developed over time for the service, leveraging the ability for digital transmission to carry a wide range of content types.

HD RADIO STANDARDIZATION

IBOC broadcasting is unique in several respects. Perhaps most unusual—and sometimes controversial—is IBOC's development and management by a single commercial organization, iBiquity Digital Corporation. Like any broadcasting format, the FCC specifies the type of transmissions that are allowed to be aired by broadcasters, and it has thus mandated that IBOC shall be the only form of digital radio broadcasting in the U.S. Normally, however, such rules direct the industry to seek technology available from any provider that wishes to participate in the standardized, open marketplace, rather than to a technology controlled by a single corporate entity. To counter this, iBiquity Digital has pledged to offer its technology on reasonable and non-discriminatory (RAND) terms to any entity that wishes to license it for the manufacture of IBOC transmission or receiving equipment.

The FCC also requested that the technology be properly standardized, which resulted in iBiquity's working with the NRSC to develop the *NRSC-5* standard, which covers all of the AM- and FM-IBOC system except for the HDC audio coding format (it remains proprietary to iBiquity). Again, however, iBiquity has pledged to license HDC to any applicant on RAND terms.

Another unique element of HD Radio is its "software-style" licensing to broadcasters. Typically, broadcasters buy equipment from manufacturers and simply use it. The ownership of any FCC-approved transmission hardware brings with it the ability to put it on the air without further negotiation by broadcasters, provided they are properly licensed by the FCC to operate it. (In other words, the cost of licensing any intellectual property used in the manufacture of the equipment is assumed to have been paid by the manufacturer, and this licensing cost is built into the price that the broadcaster pays when the equipment is purchased.) In the IBOC case, however, a broadcaster cannot legally put IBOC equipment on the air without first obtaining an additional usage license from iBiquity Digital Corporation. The terms of the license involve both one-time fees per station, as well as certain ongoing fees based on percentage of revenues arising from use of IBOC technology earned by the broadcaster (e.g., advertising on multicast channels, or datacasting carriage revenues). This unusual arrangement has caused concern among some broadcasters, which in some cases has led to their unwillingness to implement the system.

On the other hand, the system's management by a single commercial entity has allowed relatively rapid development of improvements to the system, as well as focused evangelism for the system's deployment into receivers (particularly in the automotive sector). The latter effort is particularly important for success of a system like IBOC, the adoption of which is wholly voluntary by broadcasters, with no regulatory mandate for its use and no foreseeable analog switch-off date—quite unlike the digital <u>television</u> transition, as explained in Chapter 16.

CHAPTER 14

Alternate Radio Delivery Systems

As discussed elsewhere in this book, radio broadcasting involves two basic processes—content <u>creation</u> and content <u>delivery</u>. Traditionally, the delivery component implied only over-the-air transmission via AM or FM broadcasting, but this is no longer the case. Thanks to the pervasiveness of the Internet, radio broadcasters now have other methods of delivering their content to listeners, as this chapter explains. The expansion of <u>wireless</u> broadband Internet access has had a significant impact here, placing these alternate delivery formats closer to parity with the traditionally high mobility of radio broadcasting.

INTERNET RADIO STREAMING

Many radio stations employ streaming audio services to provide a simulcast of their over-the-air signals to online listeners. In some cases, a broadcaster also may offer additional online audio streams that are repurposed, time-shifted, or wholly different from their on-air services. Because there is no scarcity of bandwidth or requirement for licensing of online services, broadcasters (and others) may offer as many streaming media services as they wish. Unlike over-the-air broadcasting, however, online distribution is delivered to end-users not by the broadcasters themselves to their local markets, but by third-party telecommunications providers on a national or worldwide basis. In addition, the cost to broadcasters of providing such signal delivery is a variable one, determined largely by the number of listeners currently streaming the service (i.e., the more listeners, the higher the cost of distribution). These are fundamental changes to the business model of traditional broadcasting, which is predicated on fixed cost of service regardless of audience size, to a specific geographic region.

There are other important differences between on-air and online services, which are largely non-technical and more business related, although they may have technical consequences. These include contractual differences between on-air and online distribution paths with programming suppliers (e.g., content networks and syndicators) or labor unions, and variations in licensing regulations affecting music royalty payments for online vs. broadcast services. The technical impact of these differences often results in the need for automated playout systems to insert different elements in a content stream (e.g., replacing certain advertisements or program segments of the on-air service in the online version), and the reporting of users logged on to an Internet radio stream during the streaming of any particular song, for purposes of determining online royalty assessments.

Streaming Media Technology

Provision of an Internet radio service is accomplished via a few basic processes, as shown in Figure 14.1. First, the audio service is assembled, and audio processing is applied. (Even if the content of the online service is wholly identical to the station's on-air service, the station may wish to process it independently.) The online audio service is then sent to a streaming media encoder—either a PC or a dedicated hardware device—whereupon one of several popular data compression algorithms (i.e., lossy codecs using "perceptual audio coding," as explained in Chapters 5 and 6) is applied to the audio program.

The compressed digital audio signal is then formatted into IP packets for streaming Internet delivery, again using one of a few different streaming media formats. Metadata is often added to the stream at this point, to provide users with text or graphics associated with the audio content, to be displayed on their devices' screens. In some cases, a broadcaster may perform this stream-creation process in several versions, each using different codecs, and/or different data rates, to serve a wider range of users.

Next, the compressed and formatted bitstream is sent to one or more *hosting servers*, where the stream is replicated and sent to listeners who have requested it. This process is usually performed by distributing multiple independent streams, one to each user—a method known as *IP unicasting*, or simply *unicast* (see Chapter 6). It is often provided to broadcasters by a contracted third party referred to as a *Content Delivery Network* (CDN), which operates at offsite locations from the station. Alternatively, a station can host its streaming services on its own servers, or work with its local Internet service provider (ISP) to do so.

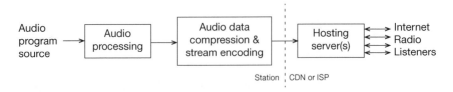

FIG. 14.1. Typical Internet Radio Signal Path

The station registers the website address—known as a Uniform Resource Locator (URL)—that it wishes to use for an Internet radio stream with an Internet name registration service, and provides that URL to the CDN or ISP so it can direct users requesting it to the appropriate server(s) to access the stream. Stations typically announce these URLs over the air, and place links to their streams' URLs on their websites, or wherever else they wish to promote their streams.

Listeners can navigate to any streaming-media URL via the web browser in their fixed or mobile device with wired or wireless Internet access, and assuming their device's media player is equipped with the appropriate software and audio codecs, they can listen to the stream's content.

Aggregation Sites

Often stations may elect to also provide their streams' URLs to one or more Internet radio *aggregation sites*. These are third-party websites that present a large number of radio stations' streams, possibly along with other non-broadcast Internet radio streams, in a single database, and offer various forms of navigation that allow users to find a particular radio service stream of interest to them (e.g., sorting by genre/format, location, program name, etc.). This allows users to store, set as a browser "favorite" or "bookmark," or otherwise recall the aggregation site's single URL, and then find the Internet radio service they are looking for using the aggregation site's navigation interface. (Some large radio broadcasting companies have developed their own aggregation sites, presenting their stations'—and in some cases, other broadcasters'—streams together in a single user interface.) The convenience of an aggregation site avoids the user having to store numerous radio stream URLs individually in their browsers, and provides a method of navigating through the many thousands of Internet radio streaming services that are available worldwide.

Interactive Streaming

Because Internet radio is delivered via a bidirectional (and typically unicast) connection, the streaming server can process responses from each user. One possible type of response from the user is whether they like or dislike the current song. This can allow the Internet radio provider to tailor future offerings. In cases where the stream is not simply a simulcast of a broadcast radio service, but is in fact an Internet-only service, the server providing the feed to the user can allow the user to skip the song and move ahead to the next. The server can also store individual users' preference information and apply it to the selection of similar songs in an attempt to maximize a customized stream to a user's taste. By leveraging this ability to gauge user response, such approaches maximize the use of the unicast transmission model in Internet radio service, which otherwise remain an inefficient and potentially costly method of reaching many users with the <u>same,</u> non-interactive streaming programs (compared to over-the-air broadcasting).

While this technology has been used primarily by non-broadcast Internet radio providers, at this writing some <u>broadcasters</u> are also offering interactive streaming services, along with the many non-interactive streams they provide, which are largely simulcasts of over-the-air broadcasts.

These value-added interactive services can be advertiser-supported, or sub-scription based. In some cases, a two-tiered service model is offered, whereby a user can receive the basic service for free with ads, or can pay a subscription fee to reduce or eliminate ads, and/or enable additional features.

AUDIO PODCASTING

Some radio broadcasters use audio *podcasts* as a method of distributing discrete, typically short-form programs, often posted shortly after the program runs on the air. This is a convenient form of "on-demand radio," which allows listeners to download programs they missed when they ran on air. Podcast-aggregation sites also offer users the ability to subscribe to future podcasts of a given radio series, so successive programs will automatically be downloaded to the user's podcasting client or device when available. This allows broadcasters to build strong loyalty to a program. Sharing or recommending podcasts provides an easy way for listeners to introduce a radio program to others, even if they are in locations where the program is not available over the air.

MOBILE RADIO "APPS"

Many radio stations have developed—often through third parties—mobile platform applications or *"apps"* for various mobile platforms that users can download (usually for free) to their mobile devices. These apps offer a variety of functions, ranging from simply directing the mobile device's browser to the station's online radio stream(s), to providing links to archival content, to presenting background data about the station and its staff, or a variety of other features related to the radio station and its content. In some cases, video ele-ments can also be offered via a radio-station app. These apps provide another way to build loyalty for a station or program.

CONVERGED RECEIVERS: "CONNECTED CARS" AND RADIOS IN SMARTPHONES

While receivers and the Internet both enjoy very high penetration, the capacity to connect to both services has rarely appeared on the same device in the U.S.

This trend is changing, however, as wireless Internet access becomes available in vehicles, and radio tuners appear in an increasing number of smartphones.

"Connected Cars"

Connected cars is the term generally applied to vehicles with wireless Internet access. In some cases, a 3G or 4G wireless data receiver is built into the car, and used for *telematics* by the car (i.e., the ability for the car to upload and download data to/from its manufacturer, for purposes of reporting status of the car, receiving updates and messages about service, downloading new versions of software, etc.). In such cases, the car may also offer a built-in browser for users to access a constrained number of online services (a so-called *"walled garden"*), including in some cases radio streaming and stream-aggregation sites. This is called an "embedded" connection.

In other cases, vehicles may offer methods of connecting their driver's or a passenger's smartphone to the car's built-in controls, display screen and entertainment system, allowing the car to use an existing external device (and its wireless broadband data service) for such online communications, and enabling the operation of the device's browser—or certain mobile apps—by the vehicle's driver or passengers with relative safety via hardware controls on the vehicle's steering wheel or dashboard, or via voice activation. This is called a "tethered" connection. With this type of architecture, the car can be connected to nearly <u>any</u> site that the user's smartphone browser or supported app can wirelessly access from a given location, which could include any Internet radio stream available.

Radios in Smartphones

In the U.S., it has been uncommon for wireless phones to include a working radio receiver (whereas such receivers have been commonplace in cell phones for many years in other regions of the world). In some cases, wireless phones sold in the U.S. actually do include radio reception capability within their hardware, but it is simply not activated on the phone's software.

Almost all of these radio tuners in wireless phones are FM only, because AM radio broadcasts are difficult to receive on such small devices (given their limited space for the larger antennas required by the longer MF wavelengths used by AM radio), and in the presence of so much RF energy inside the device.

Most FM tuners in wireless telephones use the wires of the device's headphones as an antenna. Thus FM reception is usually only possible on these devices when the headphones are plugged in. Battery drain on the device is substantially less when listening to an FM service over the air than when listening to the same service streamed via wireless Internet access to the device's browser or media player. Listening via FM also has no impact on a user's wireless data consumption, whereas streaming the same service via wireless broadband is billed against the user's data plan, typically at a rate of around 0.5 MB to 2 MB per minute for music services, and about half that for voice-dominated services such as news or sports.

HYBRID RADIO

When a user's device, vehicle, or platform has both a radio receiver and an Internet connection, those two content pathways can be integrated such that certain multimedia content elements could be delivered to the user via <u>both</u> paths in a synchronized fashion. This is called *hybrid radio*, in which the broadcast path allows delivery of a common element to all listeners (e.g., the main program audio, such as a song, or the play-by-play coverage of a sporting event), while the Internet path allows a variety of enhancements to the common content (e.g., text or graphical display of song lyrics, or listings of team statistics) that can be selected for visual presentation on the device's screen if the user so desires, creating the option for a personalized overall experience of the content. Similar arrangements can be offered for advertising, allowing audio ads delivered over the air to be enhanced by optional visual content delivered via the Internet.

Hybrid radio also allows direct response from the listener to certain elements of interactive content presented via the (inherently bidirectional) Internet path, offering a variety of new business models to radio broadcasters. Examples of such interactions include "tell me more" features (e.g., links to a website where additional information about the subject of the radio content is provided), enhanced engagement with advertisers (e.g., the location of the advertiser's nearest local outlet, based on GPS or other location-aware capabilities of the user's device), or details about music played on air (e.g., links to CD or download purchases).

A standard called *RadioDNS* has been developed that allows a receiver to use the information it already knows about a radio station (e.g., the frequency of the tuned station, and service metadata supplied by the station, such as RDS from an analog FM station), along with the receiver's general location (i.e., the country), to generate a unique URL for an associated online address. The station can then supply an enhancement stream to the station's on-air audio content at this URL (or redirect from this URL to the enhancement stream), and a rich, interactive hybrid radio service can result. Internet-connected devices with FM tuners that are equipped with RadioDNS capability can provide this service natively, or an app with tuner control can manage the hybrid radio capability and its display on the platform that RadioDNS enables. Figure 14.2 shows an example of such an app-based hybrid radio service on a smartphone.

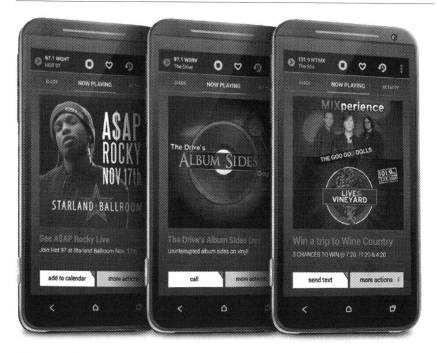

FIG. 14.2.   The *NextRadio* Mobile App uses Hybrid Radio to Provide Rich and Interactive Enhancements to Over-the-Air Content. (Courtesy TagStation and NextRadio)

As more devices with both radio receivers and wireless Internet connectivity proliferate, hybrid radio service may emerge as an important new component of the radio broadcasting industry.

AUDIO-ONLY SERVICE VIA DTV

Cable and DBS television services have offered audio-only, radio-like services to customers' homes for many years. While these are limited to listening in fixed locations (and are often more popular in other regions than they are in the U.S.), digital over-the-air television broadcasting can also carry audio-only services, including in some cases their delivery to mobile devices (see Chapters 16 for Mobile DTV in the ATSC system). It is expected that future television broadcast systems may expand this capability, particularly for mobile reception (see Chapter 18).

CHAPTER 15

NTSC Analog Television

For many years, NTSC was the system specified in the FCC rules for television broadcasting in the United States. On June 12, 2009, however, NTSC broadcasting ceased for all full-power U.S. television stations, now replaced by digital ATSC broadcasting (see Chapter 16). At the time of this writing, NTSC is still used for full-power broadcasting in Canada, Mexico, and other countries. NTSC signal outputs are also provided on set-top boxes used to convert digital TV signals for use with existing analog TV sets and other video devices.

NTSC television signals use a combination of AM and FM modulation, with subcarriers for certain parts of the signal. In Chapter 12, these methods were covered as they apply to radio transmission. If you understand the concepts discussed there, understanding analog TV transmission is quite straightforward.

CARRIERS AND CHANNELS FOR ANALOG TV

In the United States, TV channels 2 to 13 are assigned in the frequency range 54 to 216 MHz in the VHF band. TV channels 14 to 51 are assigned in the range 470 to 698 MHz in the UHF band. Prior to June 2009, the UHF frequencies 698 MHz to 806 MHz were also used for television channels 52 to 69, but that spectrum is no longer available for either analog or digital television broadcasting.

Each television channel in the United States (whether analog or digital) is 6 MHz wide. With analog NTSC television, the *visual carrier* (also known as the video or picture carrier) frequency is 1.25 MHz above the lower boundary of the channel, and the *aural carrier* (also known as the audio or sound carrier) center frequency is 4.5 MHz higher than the visual carrier. The color subcarrier is placed at 3.58 MHz above the visual carrier—all as shown in Figure 15.1.

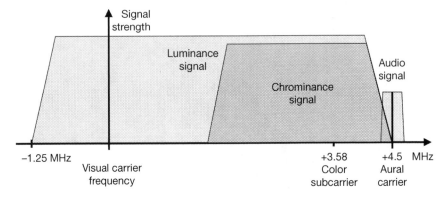

FIG. 15.1. NTSC RF Channel

Sidebands and Bandwidth

In order to conserve radio spectrum, a form of AM modulation, known as *vestigial sideband*, is used for the visual signal. This means that most of the lower sideband of the TV visual signal is cut off, as shown in the figure. As explained in Chapter 7, the normal *double sideband* signal would extend from the video carrier minus the modulating frequency to the video carrier plus the modulating frequency.

VIDEO SIGNAL

Video Carrier

The NTSC video signal is amplitude modulated onto the video carrier. The process is similar to an AM radio station—except that the TV station is transmitting video and audio information, whereas the AM radio station is transmitting audio information only. In addition, the modulated AM picture signal occupies nearly 300 times as much of the radio frequency spectrum as a modulated AM radio signal. Although the lower video sideband is curtailed, the signal carries enough information to allow the receiver to recover the video picture; the demodulation process is slightly more complex than it would be with two full sidebands available.

Color Subcarrier

In Chapter 4, we explained how the NTSC chrominance information is carried on a subcarrier within the video signal. The way this works is similar to the subcarriers for stereo and ancillary audio services on FM radio described in Chapter 12, with three main differences. The first is that this color coding is usually carried out early on in an analog system chain (for example, at the studio camera), in order to allow single-wire video distribution. Also, to allow the two color difference signals to be carried, the color subcarrier is modulated using QAM (*quadrature amplitude modulation*). The third difference is that the subcarrier frequency is within the frequency range of the video luminance information. Although the receiver does its best to separate out the luminance and chrominance, there can be some interference of one with the other, called *cross-color*. This is why with NTSC television, pictures with fine patterns, such as a check or striped jacket, can at times produce shimmering colored interference over the pattern. In that case, the particular brightness or luminance frequencies are being interpreted by the receiver as false color information.

Just as a monaural radio ignores the stereo subcarrier and uses only the main (mono) program information, a black-and-white television ignores the color subcarrier and uses only the luminance information to produce the mono-chrome picture.

The reason that chrominance and luminance share some of the same frequencies is that, when color television was first introduced, the color subcarrier had to be placed within the existing video bandwidth so that the existing TV channel frequency allocations (initially made for black and white NTSC broadcasting) could be maintained. Otherwise, all stations would have needed new carrier frequencies, and existing receivers would have been made obsolete.

AUDIO SIGNAL

The NTSC audio signal is frequency modulated onto the audio carrier at the top end of the television channel, as shown in Figure 15.1. The audio baseband signal may be a single mono audio channel or may include subcarriers for stereo and additional services, as discussed below.

Audio Subcarriers

The composite baseband signal of an NTSC television station's audio channel looks very much like that of an FM station, but with the stereo pilot frequency at 15.734 kHz instead of 19 kHz, as specified in the *BTSC stereo TV audio standard* (see Chapter 12). In addition to the stereo signal, two standardized but optional subcarrier channels are defined that can be carried with the TV audio signal: the separate audio program (SAP) channel and the professional (PRO) channel. The SAP channel is a mono audio channel that television broadcasters often used for a second language track). The PRO channel is a narrower channel than the SAP channel, and therefore only permits voice transmissions or data information, such as communications with station staff at remote locations (for example, for news gathering). An illustration of a typical NTSC TV station's aural baseband signal is provided in Figure 15.2.

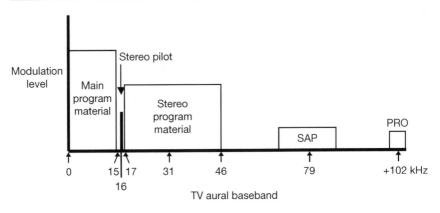

FIG. 15.2. NTSC TV Aural Baseband Signal

VERTICAL BLANKING INTERVAL (VBI) ANCILLARY INFORMATION

Because there is no picture information on the VBI lines (see Chapter 4), some of them can be used to carry additional, nonvideo information. The FCC rules specify what type of information can be transmitted on various VBI lines in NTSC. A summary is as follows:

Lines 1 to 9: *Vertical synchronization information only*
Lines 10 to 18: *Test, cue, and control signals; telecommunications*
Line 19: *Ghost-canceling reference signal only*
Line 20: *Test, cue, and control signals; telecommunications*
Line 21: *Closed captioning and other program-related information only*

CLOSED CAPTIONING AND CONTENT ADVISORY RATINGS

The FCC rules require broadcasters in the United States to carry closed captioning information for a certain percentage of most types of programming. This allows receivers to decode and display captions on screen, so hearing-impaired viewers can follow and understand the program dialogue and narration. Captioning text for NTSC video is coded into digital form at the studio and carried as a low-bit rate data signal inserted onto VBI line 21. This data is then modulated onto the visual carrier along with the rest of the video signal.

Line 21 can also carry other data associated with the program, in particular the *content advisory* information. This so-called V-chip rating can be displayed on screen and allows parents to control what types of television programs can be viewed by children. The system for closed captioning and other information carried in line 21 is specified in the Consumer Electronics Association (CEA) standard CEA-608-E.

ANALOG TV RECEIVER

The following presents a brief look at how the over-the-air NTSC TV signal is received.

A simplified block diagram of an analog television receiver, usually known as a TV set, is shown in Figure 15.3. An antenna picks up the over-the-air radio waves and feeds the radio frequency signal to the first stage of the receiver, where the RF tuner amplifies the weak signal and selects the desired broadcast channel. This signal is then converted to a lower *intermediate frequency* (IF) which is further amplified and then *demodulated* in the detector stage, reversing the amplitude modulation process described in Chapter 7. This recreates the composite NTSC video signal and FM sound signal. The composite video decoder separates the luminance and chrominance signals and demodulates the chrominance into two color difference signals. By adding and subtracting the luminance and color difference values, the receiver produces the separate red, green, and blue signals. These RGB analog video signals are amplified to drive the display device (e.g., a cathode ray tube, or an LCD or plasma flat panel display) to produce the red, green, and blue parts of the picture, which are transmitted line-by-line and

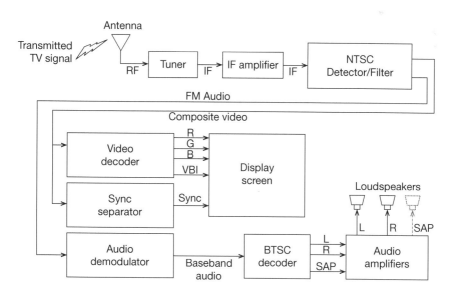

FIG. 15.3. NTSC Analog TV Receiver Block Diagram

frame-by-frame. The human eye then combines the red, green, and blue images together to see a complete, full-color picture.

Synchronizing pulses from the composite video signal are separated in the sync separator and used to produce the signals that enable the display device to show the picture lines and frames at the right time. In TV sets with CRT tubes, the synchronizing circuits produce vertical and horizontal scanning signals that control the position of the electron beam, which stimulates the display phosphors to generate light. The horizontal scan circuits also drive a high voltage generator for the extra high tension (EHT) voltage needed to accelerate the electron beam.

Where appropriate, closed captioning data is extracted from VBI lines in the composite video signal, decoded, and combined with the video signals sent to the display.

The FM audio in the IF signal is usually extracted with a process called *intercarrier sound* and then demodulated in the sound detector stage, reversing the frequency modulation process to recreate the audio signal. This may be further processed by a *BTSC decoder* to extract the left and right stereo signals, or a second audio program (SAP), and then amplified and sent to loudspeakers.

CHAPTER 16

ATSC Digital Television

ATSC stands for the *Advanced Television Systems Committee*, which developed the standards for digital television broadcasting (DTV) used in the United States (as well as in Canada, Mexico, Korea, Honduras and El Salvador, at this writing).

DTV transmissions are very different from analog TV, both in the type of video and audio signals and in the way they are modulated onto the RF carrier. ATSC DTV signals are not compatible with NTSC and need a different type of receiver—either an integrated television or a set-top box. It is possible to use a DTV set-top box to feed an existing NTSC television set, but to take advantage of the improved video quality with high definition and widescreen images, an HD television is required. The same applies to the ATSC standard's sound system—an existing NTSC television set can reproduce mono or stereo signals from ATSC signals received via a set-top box, but to receive the 5.1-channel surround sound that the ATSC system provides, a digital audio decoder and appropriate loudspeakers are required.

A fundamental difference between digital ATSC and analog NTSC television is that NTSC transmits complete pictures, but the ATSC system effectively sends the pictures broken down into many small parts, with a set of instructions to put them back together again. This is explained later in the chapter.

ATSC AND THE FCC

The names (and identifying numbers) of the principal ATSC standards to date include the Digital Television Standard (A/53), the Digital Audio Compression Standard (A/52), and the Program and System Information Protocol Standard (A/65). These standards were adopted by the FCC as the exclusive U.S. DTV format in 1996 (in 2004 for A/65), and are part of the rules that U.S. DTV broadcasters must legally comply with. An ATSC Recommended Practice on Techniques for Establishing and Maintaining Audio Loudness for Digital Television (A/85) was incorporated in part by reference in the FCC rules in 2012, as part of the rules enforcing a U.S. law passed in 2012 called the *CALM Act* (see Chapter 9, and later in this chapter).

Other standards and recommended practices for different aspects of the system have subsequently been developed by the ATSC, but these are voluntary and have not been adopted by the FCC (although the FCC rules have been revised from time to time to reference updates to the original standards made by the ATSC). The ATSC standard for Mobile DTV (A/153) is covered at the end of this chapter.

Such differences between standards and regulation are important to understand. A standard may be developed and used in multiple countries or even worldwide, but each county's government must individually adopt the standard's use as mandatory in its laws or regulations. Thus the U.S. has adopted some (or some parts) of the technical standards and recommended practices developed by the ATSC as mandatory via regulations issued by the FCC. Other ATSC standards and recommended practices may be voluntarily adopted by industry agreement within the broadcast and/or consumer electronics industries, within a country, a region or worldwide.

THE U.S. DIGITAL TV TRANSITION

The transition from analog to digital television broadcasting in the United States was a mandatory process for broadcasters, consumer equipment manufacturers, and indirectly, for consumers themselves. It certainly did not happen overnight, however. It was planned as a gradual process over many years to give broadcasters time to install new equipment, for manufacturers to develop new receivers, and for consumers to buy and install them. There were also some course-corrections and changes to the transition plan made along the way, as explained below.

DTV Transmissions

After the introduction of the ATSC standard in 1996, the FCC designed a transition plan that set a number of deadlines for different categories of broadcasters to commence DTV transmissions, starting in May 1999 with network-affiliated commercial broadcasters in the top 10 markets. By May 2003, all U.S. full-power broadcasters—commercial and noncommercial—were required to start DTV transmissions as simulcasts of their NTSC service. To make this possible, each NTSC station was temporarily assigned a new, second channel for its DTV service, using channels that were vacant in the NTSC channel allocation plan. This "table of allotments" was planned to ensure that stations could achieve equivalent coverage to their NTSC service without interfering with other stations. There were more vacant channels in the UHF band than in the VHF band, so most DTV station channels were initially in the UHF band. Because digital transmissions are more power efficient than analog, the DTV signals could achieve equivalent coverage at lower power, and thus the DTV transmissions could occupy vacant channels that could not be used by NTSC transmissions without causing undue interference. Ultimately, after a transition period where analog and digital broadcasts took place in parallel, the analog service was terminated, and broadcasters returned to single-channel operations. Full-power U.S. broadcasters completed this process in June 2009.

The FCC rules for DTV did not, and still do not, mandate that broadcasters transmit in high definition television (HDTV). Broadcasters are only required to transmit at least one program of quality equivalent to NTSC. In fact, however, most (but not all) U.S. broadcasters installed DTV equipment able to transmit HDTV programs during the DTV transition, and since then there has been a steady migration to more HDTV equipment and more HDTV programs over the U.S. DTV airwaves, though some SD programming is also still carried.

DTV Receivers

Consumer manufacturers started to introduce HDTV receivers soon after the ATSC standard was adopted, but there was not a large choice of models at first. To make sure there were suitable DTV receivers of all sizes and price ranges for consumers to buy, the FCC issued mandates for new television receivers to have ATSC DTV tuners, starting in 2005 with large-screen televisions over 36 inches. By March 2007 all new televisions regardless of screen size, and all interface devices that included a tuner (e.g., set-top box, VCR, DVD player/recorders, DVR, etc.) had to include a built-in ATSC DTV tuner.

Toward the end of the DTV transition, the U.S. government introduced a coupon program that subsidized the purchase of converter boxes to enable people to use NTSC televisions to watch DTV programs. This policy was intended to cover the last remaining segment of the market who chose not to purchase (or could not afford) new ATSC-capable TV receivers, helping to prevent them losing access to local television service over-the-air when NTSC broadcasting was shut down. The government offered U.S. consumers upon request two coupons, each worth $40 toward the purchase of a DTV converter box (one coupon per box). The program also specified what these low-cost boxes had to provide in order to be eligible for subsidized purchase with the coupons. This included the ability to receive both HD and SD DTV broadcasts and convert them to an NTSC signal that could be viewed on an NTSC set, as well as stereo sound and other features such as closed captioning, but there was no requirement for HDTV or surround sound outputs.

Spectrum Recovery Plan

In planning the DTV transition, it was assumed that NTSC transmissions would eventually be shut off and replaced by equivalent or better ATSC services. DTV enables improved picture and sound quality and new types of broadcast services, but it also uses spectrum more efficiently. It was expected that, with careful allocation of channels, TV stations could serve their existing markets using less broadcast spectrum (and power), and with less adjacent-channel interference protection than had been used for NTSC, thus freeing up valuable spectrum for other purposes. The FCC therefore decided to move all TV broadcasters previously using channels 52 to 69 into what were called the "core" channels of 2 to 51.

This plan enabled the government to reallocate the UHF spectrum with frequencies from 698 MHz to 806 MHz (formerly occupied by TV channels 52 to 69) for much-needed emergency communication services, and to auction the remainder to commercial companies for new services such as cellular telephones and broadband wireless communication networks.

As the final stage of the DTV transition, most stations were given the choice as to whether to keep their DTV transmitter on their newly assigned DTV channel, or to move the DTV service to their original NTSC channel, or in some cases, to some other channel that might have become available in that location. Stations on channels 52 to 69 (both analog and digital) were required to move to an "in-core" channel. Some stations with UHF DTV channels decided to move back to their original NTSC channel on VHF (and others did the opposite), with most stations opting to stay in the UHF band.

Analog Switch-off

The planned cut-off date for analog NTSC broadcasting was delayed several times, but eventually on June 12, 2009, all full-power NTSC broadcast stations closed down. When this occurred, U.S. television broadcasters were providing the same services (or more, if multicast services are included), at equal or higher quality, using only about two-thirds of the spectrum they had occupied using NTSC. The FCC then conducted spectrum auctions (see Chapter 20) with most of the 108 MHz of freed-up spectrum, raising approximately $20 billion.

DTV SYSTEM

A basic block diagram representation of the system is shown in Figure 16.1. According to this model, the digital television system can be seen to consist of subsystems for source coding and compression, service multiplex and transport and RF/transmission, all of which are explained in this chapter.

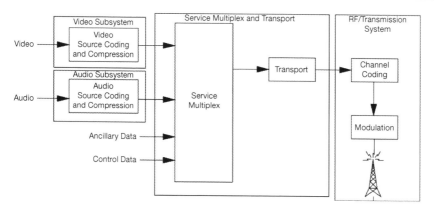

FIG. 16.1.   ATSC System Block Diagram. Courtesy of ATSC

(Upstream elements of digital television audio and video production that inter-face with ATSC DTV transmission are covered in Chapter 9.)

CARRIERS AND CHANNELS FOR DTV

DTV RF channels are in the same VHF and UHF bands as originally used for NTSC television (although the UHF band is now smaller, as explained above). In the United States, TV channels 2 to 13 are assigned in the frequency range 54 to 216 MHz in the VHF band. TV channels 14 to 51 are assigned in the range 470 to 698 MHz in the UHF band. Not all frequencies within in these ranges are assigned to television, however; there are gaps for other services (e.g., the FM radio band at 88 to 108 MHz, and radio astronomy on UHF TV channel 37).

As with NTSC, each DTV channel in the United States is 6 MHz wide. There is one carrier for each 6 MHz channel in the ATSC system. A reduced-level carrier frequency, known as the *pilot*, is placed at the lower edge of the channel; the receiver needs this pilot carrier to help lock onto the incoming signal. Figure 16.2 shows a plot of the spectrum of an actual ATSC transmission channel.

Channel Navigation and Branding

To make it easy for consumers during the DTV transition, the ATSC stand-ard included a feature that allowed a DTV broadcast station to use its previous NTSC channel number as the *major channel* number for the DTV service, com-bined with a second *minor channel* number. The DTV major channel number helped the station retain its brand image developed during its years of provid-ing analog service to its community. This identifying channel number could

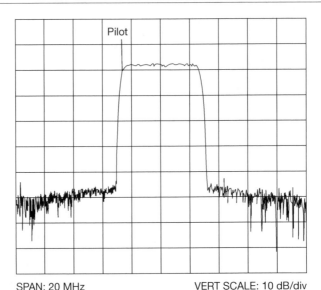

SPAN: 20 MHz   VERT SCALE: 10 dB/div

FIG. 16.2. ATSC RF Channel

remain constant, even if the station moved its actual DTV RF channel during the transition period. Thus today there are many DTV stations that are identified by a channel number that is different from the actual TV channel number on which they broadcast. For example, in Washington, DC, WRC-TV continues to identify itself as Channel 4, but actually broadcasts on TV channel 48 (674–680 MHz). See the section on PSIP below for more details.

Sidebands and Bandwidth

The eight-level vestigial sideband (8-VSB, see below) modulation used by the ATSC format inherently produces numerous sidebands above and below the carrier frequency. To keep within the 6 MHz channel bandwidth, the upper sidebands are greatly reduced, and virtually all of the lower sidebands are cut off, leaving only vestigial sidebands.

Nevertheless, the reduced signal carries enough information to allow the data payload to be fully recovered by the receiver.

8-VSB MODULATION

Digital Modulation

8-VSB is a special form of AM modulation using a single carrier modulated to eight different amplitude levels. The carrier is modulated with a compressed bitstream signal that, as explained later, carries all the information needed for the channel. This modulation occurs at the transmitter in a piece of equipment called an *exciter* (see Chapter 19 for more on exciters). The compressed bitstream input to the DTV exciter is processed and converted into digital *symbols*. The output of the exciter is a series of pulses of carrier wave, each with one of eight different amplitude levels that are assigned to eight symbols. These symbols are output at a rate of 10.76 million every second.

Transmitted Bit Rate

Each symbol in an 8-level modulation system is able to carry three bits of digital data. Therefore, the 10.76 million symbol-per-second output of a DTV exciter can represent $10.76 \times 3 = 32.28$ Mbps (megabits per second) of digital data. It is important to understand that the "data" being transmitted may represent video, audio, or other types of ancillary data information. Once in the bitstream, it is all referred to as data.

As explained in the later sections on compression and transport streams, however, the information (or "payload") bit rate for an ATSC bitstream is only 19.39 Mbps. This is the <u>net</u> bitrate available for audio and video programs and other broadcast content. The remaining data from the calculation above—nearly 13 Mbps—is used for *forward error correction*.

Forward Error Correction

Forward error correction (FEC) is used to make the signal more robust and able to withstand noise and distortions that will naturally occur in any wireless transmission path. Errors in the received bitstream may be caused by practical limitations in the transmitter and receiver, by variations in radio wave propagation, or by interference from other stations on the same or adjacent channels. There may also be various forms of interference such as lightning strikes that create large RF noise pulses, which can affect reception at great distances from the strike. Smaller RF pulses are created by electric motors, car ignition systems, and many other things. All of these conditions can cause data to be corrupted or lost. The special error correction codes added to the DTV signal help receivers to detect and correct problems with the received signal.

Adaptive Equalizer

Multipath distortion is another common problem with over-the-air television signals, caused when multiple signals from the same transmitter arrive at the receiver at slightly different times due to reflections (see Chapter 20 for more on information on multipath). Such distortions, also known as *echoes*, can result in loss of received data in DTV receivers. To counter this, DTV receivers have special circuitry called *adaptive equalizers*, which help cancel out the effects of unwanted multipath signals. The adaptive equalizers in modern ATSC receivers are far better than those in early sets, and they greatly contribute to reliable DTV reception today.

Cliff Effect

With analog television, the received picture can be excellent (in areas of strong signal reception), or may be mediocre, exhibiting "ghosting" and "snow" (in areas of marginal reception), or unwatchable (in areas of poor reception). ATSC DTV pictures never display ghosts because, so long as the receiver can decode the digital data that describes the picture, it can reconstruct a virtually perfect picture. Under extreme conditions, however, with low received signal strength and/or high levels of interference from other sources and/or strong multipath echoes, the DTV receiver may no longer be able to correctly decode the picture (and/or the sound). In that case, the picture and sound may break up, freeze or disappear completely. The signals go from perfect to nothing, with almost no intermediate stage. This is known as the "cliff effect," which is common to many digital transmission technologies.

ATSC COMPRESSED BITSTREAM

The ATSC compressed bitstream, comprised of digital video, digital audio, and ancillary data, is known as a *transport stream*. This is created by the *ATSC encoders* and *multiplexer* (see Figure 9.11 in Chapter 9), and delivered to the television station's transmitter at a rate of 19.39 Mbps. The transport stream is made up of blocks of compressed digital data, known as *packets*, which carry video, audio, and ancillary data. This is in contrast to the continuous stream of analog information for an NTSC signal.

Each ATSC packet contains 188 bytes of data, but the first byte of each packet is used for synchronization, and an additional three packets are used for management of the bitstream, so 184 bytes are actually available for audio, video, data, and PSIP (see Figure 16.3).

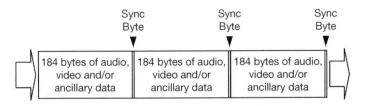

FIG. 16.3.   ATSC Data Stream

ATSC VIDEO FORMATS

Most of the characteristics of digital video signals used for ATSC transmission are discussed in Chapter 5. The signal that is actually routed around the television studio and used for editing, recording, and so on, can be one of those various video formats—and in fact, some stations continued to use their NTSC analog video equipment in the studio throughout the early stages of digital television implementation. In order to broadcast an ATSC DTV signal, the station must have an ATSC encoder that is capable of taking the video format used by the station's studio equipment as an input, which it then encodes for broadcast in a standardized fashion.

The video formats originally specified in the ATSC A/53 standard are listed in Table 16.1 below.

There are 18 different formats listed in the table. This becomes 36 if the same formats using the small frame rate adjustment mentioned in Chapter 6 are counted (i.e., each 60-frame format is also supported at 59.94 Hz, each 30-frame format at 29.97 Hz, and each 24-frame format at 23.976 Hz). The number of pixels and lines refer to those coded for transmission. Horizontal and vertical blanking intervals are not transmitted in DTV, hence the VBI lines used for closed captioning and other functions in NTSC cannot be used. (Instead, such content is included in the digital bitstream.) The 1080 and 720-line formats are *high definition* (HD) and the 480-line are formats are *standard definition* (SD).

You may note that the numbers of lines and pixels for the 480-line formats are slightly different from those listed as standard definition video production formats (720 × 483) in Chapter 5. A few extra lines and pixels have to be discarded so the picture can be broken down into the right size pieces for compression.

TABLE 16.1
Video Formats Supported in the ATSC A/53 Standard

Resolution (Pixels × Lines)	Aspect Ratio	Frame Rate (Progressive or Interlace)			
1920 × 1080	16:9	24p	30p	30i	
1280 × 720	16:9	24p	30p		60p
704 × 480	16:9	24p	30p	30i	60p
704 × 480	4:3	24p	30p	30i	60p
640 × 480	4:3	24p	30p	30i	60p

Also note that 704 × 480 formats use a concept called *non-square pixels* to accommodate backward compatibility with NTSC analog video content, for its proper conversion to SD digital video (i.e., 704 × 480 does not exactly correspond to a 4:3 ratio).

There is no 1920 × 1080 format at 60p (60 progressive frames per second) in the ATSC A/53 standard, because the high data rate needed to transmit this format cannot be adequately transmitted using MPEG-2 coding (see below) within the payload bit rate available in the ATSC system. The 1280 × 720 format does not include a 30i version (30 interlaced frames per second), because there is no existing 720-line interlaced video format with which compatibility has to be maintained.

The 704 × 480 interlaced 4:3 format is used to transmit programs produced as standard definition digital video, or for existing NTSC video signals converted to digital, as noted above. The 640 × 480 formats were included for compatibility with the computer VGA graphics standard, but are rarely, if ever, used for DTV.

The table of video formats shown above was the one part of the ATSC DTV standard on which all parts of the industry could not agree. Hence, when the FCC incorporated the standard into its rules, the video format table was omitted completely. Broadcasters are therefore allowed to transmit any available video format. In reality, most broadcasters use either the 1080-line 30i or 720-line 60p formats for HD content, and the 480-line 30i for SD content.

ASPECT RATIO MANAGEMENT

One of the key features of ATSC is its ability to present video at both HD and SD resolutions, which also implies managing images with aspect ratios of both 16:9 and 4:3. The system also maintains compatibility to *downconvert* all HD video for display on NTSC screens, which have a fixed 4:3 aspect ratio. To accommodate this, methods have been developed to allow the display of images of either aspect ratio on either screen, as follows. 16:9 content can be presented on 4:3 screens at either the full <u>width</u> of the 16:9 image, or the full <u>height</u> of the 16:9 image. When full width is used, the presentation is called *letterbox*, for the black bars added above and below the 16:9 image (see Figure 16.4). When full height is used, the presentation is called *center cut*, because the edges of the 16:9 image are cut off. Conversely, when 4:3 content is viewed on 16:9 screens at full height, the presentation is called *pillarbox*, for the black bars added on either side of the 4:3 image. (4:3 content cannot be shown at full width on a 16:9 screen without distorting the image—usually called "stretching"—or cutting off the top and bottom of the image, neither of which typically present an acceptable result.)

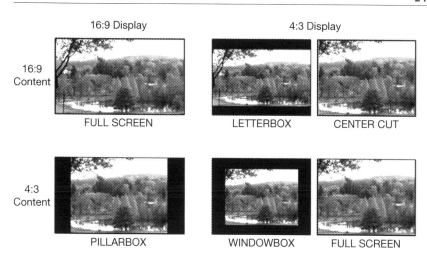

FIG. 16.4. Presentation Options (and the Common Names for Each) for 16:9 and 4:3 Content on 16:9 and 4:3 Displays

Active Format Description

In practice, it remains common for a broadcast channel to frequently switch between content of 16:9 and 4:3 during different programs, or even within a single program (such as in an advertising break). This complicates the downconversion process for the station's display on 4:3 screens. To address this issue, the *Active Format Description* was developed, which is a 4-bit metadata code that has been standardized by SMPTE, ATSC and CEA as a method of describing aspect ratio and picture characteristics of digital video signals. It is used by broadcasters in the ATSC format to dynamically control how downconversion equipment adapts 16:9 pictures for 4:3 displays. This is helpful for users with 4:3 screens if they are watching a DTV station directly off the air, but is most important for cable and satellite TV systems who must select a fixed, full-time presentation format for each DTV station they carry for presentation to SD customers.

Without AFD, either a fixed letterbox or center-cut approach must be selected by the cable or satellite system for each DTV channel (or manually selected and changed as necessary by the off-the-air 4:3 viewer—a burdensome task) for presentation of 16:9 content on 4:3 screens. Letterbox is usually preferred, so none of the image information is lost, but when the station airs 4:3 content, the fixed letterbox downconversion format results in the undesirable *windowbox* (i.e., a combination of letterbox and pillarbox, also called "postage stamp") effect, as shown in Figure 16.4. Going to a fixed center-cut mode results in loss of the 16:9 picture's edges, as mentioned, which for some HD content (e.g., an HD sports broadcast) can degrade the viewing experience, and may result in graphical elements being cut off for the 4:3 viewer. Thus there is no single, set-and-forget downconversion setting for both 16:9 and 4:3 content viewing on 4:3 displays.

AFD therefore provides the solution, allowing downconversion to dynamically adapt to an optimum presentation mode for both 16:9 and 4:3 content on 4:3 displays. It can also be set upstream by the content producer to provide the desired effect on both types of displays. For example, a 16:9 content creator may prefer to have a full-height (i.e., center-cut) picture display on 4:3 sets—even for just one particular scene within a longer program—and doesn't care if the edges of the picture are cut off for that scene. AFD can also accommodate optimal display on both screen types for content produced in other aspect ratios, such as certain widescreen movie formats. Another capability is the inclusion of *bar data*, which specifies the screen layout of the black bars in a boxing scheme, particularly useful when unusual aspect ratios or downconversions are encountered.

Pan and Scan

Another method of accommodating widescreen content on narrower display formats is called *Pan and Scan*. This is not a delivery system to end users, but a production system, which allows dynamic adjustment by the content owner or broadcaster, to determine what parts of a widescreen original production are delivered via broadcasters to users with narrower displays, with adjustments made on a scene-by-scene basis (or even within a scene, if necessary). This allows a movie director, for example, to decide how a production originally produced for widescreen cinematic presentation will be displayed on both 16:9 and 4:3 screens, with the option to adjust the framing dynamically during the course of the program, as needed to accommodate what the director feels should be in the center, on the edges, or off the screen in the narrower formats.

MPEG-2 COMPRESSION

As shown in Chapter 5, the total bit rate for uncompressed HD video is 1.485 Gbps, and for SD video 270 Mbps. These rates are far higher than the payload that can be transmitted in the 19.39 Mbps ATSC bitstream; therefore, it is necessary to first reduce the rate using data compression (the basic operation of which is also discussed in Chapter 5).

The ATSC standard is based on using *MPEG-2 video compression*. MPEG is the Moving Pictures Experts Group, which has developed various standards for compressing digital video, and is part of the International Organization for Standards (ISO). As in all forms of digital broadcasting, data compression is an extremely important aspect of the ATSC system—without it, transmitting digital television would not be practical.

The perceptual coding (described in Chapter 5) that is used in the MPEG-2 system is capable of data rate reductions of around 60:1 without significantly noticeable degradation. Here are the key elements of its operation (which are similar to the workings of most other video data compression algorithms).

Discard Unneeded Data

The first basic principle is the elimination of all information that is not actually needed to produce the picture. Some of the data that can be discarded is the signal that occurs during the horizontal and vertical blanking periods, where there is no picture. The number of bits used to describe each picture sample is usually reduced from 10 to 8. In addition, the amount of picture color detail can be reduced without being noticeable to the human eye. Chapter 5 discussed color subsampling, and in the ATSC format 4:2:0 subsampling is applied. In some cases, the amount of luminance (brightness) detail may also be reduced.

Exploit Redundancy

Video frames often have a great deal of redundant information that can be reduced or eliminated to decrease the amount of data that needs to be sent. A shot of a still picture is an extreme example. Instead of transmitting a stream of identical pictures 30 times per second, the DTV system could send just one complete frame of information, and then an instruction to the receiver to repeat the frame over and over, until it changes.

Even within a single frame, there may be redundancy. For example, let's assume that a particular video image has a red horizontal line one pixel high and 100 pixels in length extending across the screen. The uncompressed video data for this image might include instructions such as the following:

> Display pixel 1 in red
> Display pixel 2 in red
> Display pixel 3 in red
> …
> Display pixel 99 in red
> Display pixel 100 in red

However, the compressed video data for this image could include instructions such as these:

> Display pixel in red
> Repeat 99 times

This greatly reduces the amount of transmitted data, while effectively sending the same information. Using this approach within a single frame is called *intraframe coding*.

Motion Estimation and Interframe Coding

Television pictures often have an object that does not change much from frame to frame, but just moves across the screen. Often, it is possible to predict where it is heading in the next frame by looking at where it came from in the previous one; this is called motion estimation, and it can be used to reduce the amount of replicated data that is transmitted across a sequence of frames. In addition, MPEG-2 limits the encoded information for some frames and sends only what has changed since the last picture. Exploiting the redundancy of information between multiple frames is called *interframe coding*, and it further reduces the amount of new information that needs to be sent to construct the next frame.

MPEG Frames

Unlike NTSC video signals where all frames were structurally identical and remained that way from camera to transmitter to display screens, ATSC video frames are of multiple types, and the frames transmitted are different from those captured by the camera or display on viewers' televisions. There are three distinct frame types produced by the MPEG-2 video compression used in ATSC: *intracoded* frames, *predictive coded* frames, and *bidirectionally predictive* coded frames. These are generally referred to as I-frames, P-frames, and B-frames, respectively.

An I-frame is intraframe-coded, using compression techniques only within its own borders. Thus an I-frame is a complete picture and can be decoded and displayed at a receiver by looking only at the data within itself. Consider Figure 16.5 as an example of an I-frame.

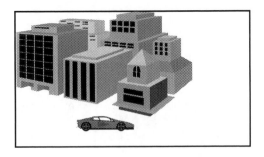

FIG. 16.5.   I-Frame

In contrast, a transmitted P-frame contains only the "change" information needed to re-create a video frame <u>in conjunction with</u> the previous I-frame that has already arrived at the receiver. Figure 16.6 illustrates the concept of a P-frame following an I-frame.

Finally, a transmitted B-frame contains the very small amount of "change" information needed to re-create a video frame when combined with information from two other frames—a preceding I-frame or P-frame and a subsequent I-frame or P-frame. Like a P-frame, a B-frame cannot stand by itself. Figure 16.7 shows the entire sequence.

The transmitted P- and B-frames always carry less new information than an I-frame, and therefore help reduce the data rate that is required to be transmitted. The number and order of P- and B-frames between two I-frames is called a *group of pictures* (GOP) and may vary for different MPEG encoders. Long GOPs have many P- and B-frames and thus reduce the data rate best. The disadvantage of long GOPs, however, is that the extended time between I-frames slows down picture acquisition when the viewer changes channels or regains signal after a

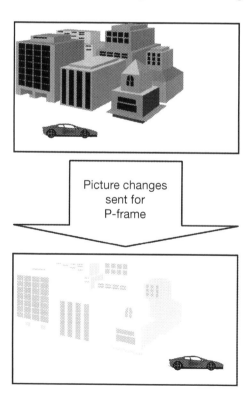

FIG. 16.6.   I-Frame Followed by P-Frame

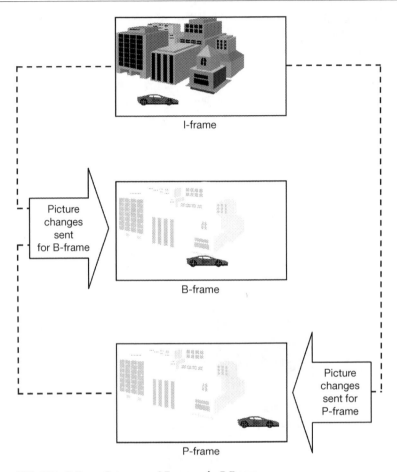

FIG. 16.7.   B-Frame Between an I-Frame and a P-Frame

dropout (an image cannot be displayed on the receiver's screen until an I-frame is received), and the GOP is also more susceptible to interference.

Decoding

Although this signal that is transmitted over the air contains enough information to re-create a video image on the television screen, it does not directly carry all of the video data for each pixel of every picture, nor does it carry them in proper viewing order. Instead, the signal received off the air effectively carries the <u>instructions</u> that enable computing circuits inside a DTV receiver to properly re-create the image. The MPEG decoder in the receiver thereby reconstructs and reorders a series of identical frames for sending to the display screen, equivalent to what was originally sent from the studio before being turned into the three different types of ATSC transmission frames and shuffled. The decoded output frames all have identical display characteristics, and are equally full of viewable content, even though some of them (P-frames, and particularly B frames) were re-created from <u>transmitted</u> frames with less information than others (I-frames).

Buffering

It may seem counterintuitive that a particular frame in a video stream could be reconstructed, in part, from a future frame that has not even made it to the screen yet. But this is possible because the ATSC video stream is not immediately displayed on the screen upon its arrival at the receiver. During the decoding process, the video data is temporarily stored, processed, and then eventually forwarded to the screen. This "store and forward" process is called *buffering*, a common practice in digital transmission. While the video data is temporarily stored in the receiver, the computer circuitry in the decoder has an opportunity to process or manipulate it. A similar buffer is used at the MPEG <u>encoder</u> at the broadcast facility, so that the order in which the frames are transmitted can be changed for the particular GOP sequence being used.

As noted earlier, a B-frame is constructed with information both from previous and from subsequent frames. For the receiver to be able to reconstruct a complete frame for display from a transmitted B-frame, the two frames from which it is predicted must be transmitted before the B-frame. For example, if three frames of video have been encoded at the studio in a sequence containing an I-frame first, a B-frame second, and finally a P-frame, the order will be changed for transmission using the buffer in the encoder. This reordering will result in the I-frame being transmitted first, followed by the P-frame and then the B-frame. The receiver can then receive and store the information for both the I-frame and the P-frame before it has to decode the B-frame, which will then be re-sequenced back into the middle position when displayed.

The buffering needed for this process is the primary reason that a brief delay occurs between the instant the viewer changes a DTV channel and when the new program appears on the screen. Also contributing to this delay is the fact that, in order to begin decoding the new video stream, the decoder must wait for the next I-frame to be transmitted for that program. This is because the I-frame is the only one that can stand by itself without any previous frames to use for prediction, so in order to begin decoding an ATSC video stream, a receiver must start with an I-frame.

Efficient Encoding

A final step in the data compression process takes place after the removal of redundant information described above. The data that is left to be transmitted can be coded efficiently, taking advantage of the variations in the makeup of the video pixel information. An example of efficient coding is Morse code, which uses combinations of dots and dashes to transmit text messages letter by letter. The letter E typically occurs most often in the English language, so that is coded as a single dot symbol. However, an infrequently used letter, such as Q, is coded with four symbols (dash, dash, dot, dash). Thus, a typical message would be sent with more short codes and less long codes, reducing the amount of data required.

Although the type of compression used in ATSC video is more sophisticated than that, the same basic principle of coding efficiency still applies, and one can see how it is possible to code raw video data in a manner that requires less data overall to transmit but does not significantly degrade the picture.

ADVANCED VIDEO CODECS

MPEG-4 Part 10 AVC

Developments in compression technology since the introduction of MPEG-2 in the mid-1990s now allow high-quality encoding with even greater efficiency. In particular, the MPEG-4 Part 10 AVC *Advanced Video Coding* system, developed by a Joint Video Team in conjunction with the ITU, where it is known as ITU-T Rec. H.264, provide equivalent picture quality at about half the bit rate of MPEG-2, or less. (The newer codec is typically referred to as "H.264," "MPEG AVC," or simply "AVC," although it is occasionally—and erroneously—called "MPEG-4" or "MPEG-4 Video".)

Introduced in 2003, AVC builds on the same principles of video compression as MPEG-2, but is able to take them to a higher order, owing mostly to the improvements and increased cost effectiveness of microprocessors and memory chips that have been developed in the interim.

Such systems allow more programs to be transmitted in less bandwidth, and the ATSC has adopted AVC as an approved alternative codec to MPEG-2, particularly for possible use in countries now introducing DTV, where maintaining compatibility with existing DTV receivers based on MPEG-2 is not an issue.

AVC coding is not compatible with MPEG-2 decoding, so in the interest of not making existing ATSC receivers obsolete overnight, U.S. broadcasters must continue to use the MPEG-2 standard. (Note that most codecs' encoder equipment performance improves over the period of the codec's usage, such that about a 30% increase in efficiency can be expected without changing decoding equipment in receivers, as encoders evolve over the years. This has been the case for MPEG-2 video encoders, as well, so even though the large installed base of legacy receiver hardware dictates remaining with the same codec, quality can still incrementally improve over time.) Meanwhile, more recently developed standards such as the Blu-ray have implemented use of AVC from the start.

Many television receivers produced at this writing also include AVC decoding capability, and some cable and satellite television systems have moved to AVC, or plan to do so. Certain newer extensions to the ATSC standard allow the use of AVC, such as Mobile DTV (see below and Chapter 18).

COMPRESSION ARTIFACTS

If insufficient bits are used to encode the video, the resulting picture exhibits degradations known as compression artifacts. The most noticeable are *pixelation* and *blocking*. When this happens, instead of natural-looking pictures, the picture breaks up instantaneously into small or larger rectangles with hard edges, either in particular areas or all over the screen. This may happen continuously or just at particular difficult events such as a dissolve, when every video pixel is changing. The AVC codec employs "deblocking filters" which help to reduce the visibility of blocking artifacts when insufficient bits are available.

No specific formula can be used to calculate exactly how many bits per second are needed to transmit a given program at a certain quality level. The data rate needed is not a constant number. It is partly dependent on the frame rate and the resolution, but it also depends very much on the subject matter of the picture. All else being equal, higher resolution, higher frame rates, and video with lots of detail and motion in it all require higher data rates.

Using MPEG-2 compression, HD video can be satisfactorily encoded at rates between about 12 and 19 Mbps, and SD video at rates from about 2 to 5 Mbps, depending on picture content. For material originally shot at 24 frames per second (film rate), the bit rate can be reduced by about 20 percent compared to 30-frame interlaced material, with similar quality. More recent implementations of MPEG-2 codecs continue a trend to improve picture quality at lower bit rates and, as mentioned previously, AVC and other advanced video codecs can reduce the bit rate by 50% or more over MPEG-2 video coding for equivalent or better quality.

AC-3 AUDIO

The digital audio that accompanies ATSC video has a very different format from that used in NTSC television or traditional radio broadcasting. In particular, it has the capability for surround sound with 5.1 channels, and it is data-compressed to reduce its bit rate for transmission.

As described in Chapter 15, in NTSC the analog video and audio signals are transmitted on two separate carriers. In ATSC, the digital transmission has a single signal that is a continuous stream of data packets. Each individual data packet can carry audio, video, and/or ancillary data. It is up to the ATSC receiver to sort them all out.

The packets of audio data in an ATSC signal conform to a system developed by Dolby Laboratories called *AC-3* (for Audio Coding 3), which is also known as *Dolby Digital.* The specifications for this system are part of the ATSC standard.

Surround Sound

AC-3 is designed to provide up to six channels of surround sound. One of these is a *low frequency effects* (LFE) channel that provides low-frequency audio to a subwoofer speaker that enhances audio effects, such as explosions or trains passing by. Because of the limited audio frequency range carried over the LFE channel, this channel requires much less data to convey its audio information than the other, "normal" or "full-range" audio channels. For this reason, the LFE channel is sometimes considered to be only one-tenth of a channel, and the overall ATSC audio system is known as a 5.1 (rather than 6) channel system.

The six channels of audio are intended to be heard through speakers that are generally positioned as shown in Figure 16.8. Because very low frequency sounds are nondirectional, the position of the low-frequency subwoofer is not critical, so this speaker can usually be positioned wherever it is most convenient.

When surround sound is not needed for a program, AC-3 can also carry regular two-channel stereo or single-channel mono signals. These are sometimes referred to in AC-3 parlance as 2.0 or 1.0 services, respectively.

Additional Audio Services

In addition to the main program audio in the principal language, the ATSC system allows for one or more additional versions of the audio track to be carried with alternative languages that can be selected at the receiver.

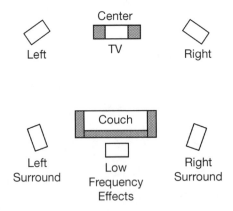

FIG. 16.8. Layout of 5.1-Channel Surround Sound Speakers

AC-3 is also designed to allow other services to be carried simultaneously, to be selected by the viewer to supplement or replace the *complete main* (CM) service. These are as follows:

- Service for the visually impaired (VI)
- Service for the hearing impaired (HI)
- Associated service—dialogue (D)
- Associated service—commentary (C)
- Associated service—voice-over (VO)
- Main service—music and effects (ME)
- Emergency information (E)

The *Visually Impaired* service is an audio channel used to describe the video scene that it accompanies. Its purpose is to allow visually impaired people to "watch" a television program by receiving periodic audio updates of the on-screen activity, in addition to hearing the complete main program audio. (This service is sometimes called *Video Description.*)

The *Hearing Impaired* service is an audio channel that contains dialogue that may be processed for improved intelligibility by hearing-impaired viewers, in addition to containing the rest of the complete main program audio. (Closed captioning information, which can also be provided for hearing-impaired viewers, is transmitted separately as video data.)

The *Dialogue, Commentary,* and *Voice-over* services were designed to be used when required with a Music and Effects service, which contains only the *Music and Effects* from the program and not any dialogue. This was intended as an alternative method to enable multiple language versions of the same program to be broadcast. Using this method, a viewer theoretically can receive the video and accompanying music and effects, and then select between the different languages available from the Dialogue, Commentary, or Voice-over services. Because of the complexity involved in creating several separate dialogue signals to accompany the same music and effects, this combination is not currently used. In any case, few, if any, DTV receivers have the two AC-3 decoders that are needed to implement this feature.

The *Emergency* (E) service is intended to allow the insertion of supplemental emergency audio announcements.

Audio Data Compression

If uncompressed digital 5.1 surround sound were transmitted, it would take up more than 4 Mbps of the 19.39 available bitstream data rate. The most important feature of the AC-3 system is that it allows high-quality data compression of the audio signals, reducing its bit rate by a factor of about 12.

AC-3, like most audio compression systems, is a *perceptual coding* system that relies on human *psychoacoustic* principles to allow certain parts of the audio signal to be discarded, or sent with less accuracy, without significantly changing the sound heard by the listener, as described in Chapter 5. Some of these principles of psychoacoustics include the following:

- Sounds are inaudible if they are below a certain level, which varies with frequency.
- Loud tones mask soft tones of a similar frequency occurring at the same time. In most cases, the human hearing system will not even recognize the existence of the quiet tone.
- Louder sounds mask softer sounds immediately after or before the occurrence of the louder sound.

Therefore, the digital bits used to represent the sounds that would not be heard by the human listener anyway can be discarded without perceptibly altering the audio. After bits have been discarded, special digital coding techniques can be used to further reduce the bit rate.

The output of the AC-3 encoder is a bitstream containing packets of compressed audio data, typically at a data rate of 384 kbps or less.

Loudness Control

Chapter 9 described the concept of audio loudness (as opposed to audio level), and introduced another key feature of the AC-3 system—its ability to maintain a consistent loudness across different segments of content, as might be often encountered in watching a broadcast TV channel (e.g., network entertainment program, advertising break, local station insertion of other ads and identification, network public service announcement, then back to the entertainment program). AC-3 does this through the use of metadata accompanying all audio data packets called *Dialog Normalization* ("dialnorm"), which lets the ATSC receiver know how perceptually loud the current content is going to sound relative to a reference value, so the receiver can adjust its sound output level to keep the perceived loudness relatively consistent. FCC rules mandate the use of this feature by broadcasters.

Ideally, the use of dialnorm would allow broadcasters to assemble streams of content from various sources without worry of inconsistencies in audio volume, and the receiver would sort it out by automatically adjusting its audio volume to keep the loudness constant. This would be independent of the user's own volume control, which can be set for a comfortable and appropriate volume in the viewing room. Whatever volume the user selected, dialnorm would keep the perceived loudness of all the content elements at approximately that same volume.

Unfortunately, the complexities of the television content ecosystem, coupled by the lack of loudness measurement standards and other difficulties kept this system from achieving this objective in all cases after the digital TV transition. The problem was worsened by the fact that the DTV audio purposely offered very wide dynamic range (much wider than NTSC audio, in which the heavy audio processing typically used by broadcasters forced all the audio of all content segments into a fairly narrow loudness range). This potentially improved the quality of the overall DTV viewing experience, but also allowed for even larger jumps in volume when things went wrong and loudness was not properly matched between program segments. Such mismatches were most annoying to viewers on transitions between programs and advertisements, with the latter sometimes playing at a substantially louder perceived volume when uncorrected in the ATSC receiver. This problem occurred when the dialnorm value in the broadcast metadata did not accurately represent the actual loudness of the sound, so the receiver would not make the appropriate adjustment, and a large jump in volume could result.

Broadcast engineers soon recognized this problem, and working with the ATSC, a *Recommended Practice* document known as ATSC *A/85* was developed, which presented several potential solutions broadcasters (and downstream cable and satellite TV providers) could employ to enable the loudness control features of the AC-3 system to work as originally intended—even within the complexities of a real-world television broadcast ecosystem.

The CALM Act

Nevertheless, this loudness mismatch problem became so pervasive after the DTV transition that an actual law was passed by the U.S Congress known as the Commercial Audio Loudness Mitigation (CALM) Act mandating that broadcasters follow the recommendations of ATSC A/85 (insofar as commercials were concerned), which, in essence, presented methods by which broadcasters could ensure that transmitted dialnorm values accurately represented the audio loudness in every content segment. See Chapter 9 for more on studio practices involved with loudness control and CALM Act compliance.

ADVANCED AUDIO CODECS

The more recent advanced audio codec systems such as ISO/MPEG-4 AAC and a later version, HE-AACv2, provide higher audio quality at lower bit rates than AC-3. For example, whereas as AC-3 is typically allocated 384 kbps for 5.1-channel surround audio and 192 kbps for stereo, HE AAC v2 can achieve comparable or better audio quality at rates of 160 kbps for 5.1 and 48 kbps for stereo.

Like MPEG-2 video coding, existing ATSC receivers use AC-3 only for audio, so broadcasters must continue to use AC-3 coding to avoid obsoleting consumer equipment. As mentioned above, however, more recent extensions of ATSC DTV may incorporate newer codecs, and HE-AACv2 is specified in ATSC Mobile DTV (see below).

AAC coding improves over earlier audio coding like AC-3 or the popular MP3 format mostly by leveraging improved and more cost-effective micro-processor capabilities, just as with the video codec improvements over time explained above. HE-AAC uses a technique called *Spectral Band Replication* (SBR) to improve high-frequency audio response in low-bit-rate applications, while HE-AAC v2 couples SBR with *Parametric Stereo* (PS) to enhance the compression efficiency of stereo signals.

MULTIPLEXING

As previously mentioned, the compressed bitstream fed to the ATSC transmitter is known as a transport stream, and contains video, audio, and ancillary data packets. The standard used for this is also defined by MPEG, and is generally referred to as the *MPEG-2 Transport Stream* (or MPEG-2 TS) standard. (It is impor-tant to distinguish between MPEG-2 TS and MPEG-2 video coding, which are two independent standards that can be used together or separately.) The com-pressed video packets from the MPEG encoder and the compressed audio pack-ets from the AC-3 encoder for a particular program are combined or *multiplexed* together to produce the transport stream. At the same time, other system infor-mation data such as *PSIP* (see below) is added to the multiplex, together with ancillary data—if any—for datacasting (see below).

The multiplexing process can be explained by thinking of the packets as rail-road cars, each containing one type of data, and the multiplexer as a freight switch yard where the various wagons are assembled into a continuous train. The *packet scheduler* permits packets to enter the transport bitstream according to need and priority. This allows dynamic allocation of the channel.

Figure 16.9 illustrates how the ATSC transport stream can be considered as a large data "pipe," which is divided up into smaller "pipes" carrying multiple types of information. Don't forget that in reality all of this information is being sent as serial data (i.e., as bits sent one after the other down a single channel).

MULTICASTING

The total bit rate for an ATSC program service comprises the audio and video bit rates plus a small amount for system information. In many cases, a single DTV program may require less than the 19.39 Mbps available for transmission in the 6 MHz ATSC DTV channel. When this occurs, it is possible to transmit multiple program services over the same DTV channel. This is done by add-ing additional video and audio compressed bitstreams to the multiplex, and is

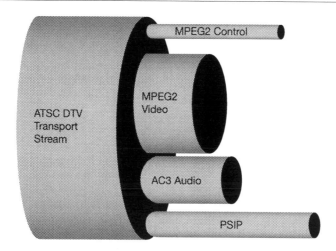

FIG. 16.9.   Data Pipe Concept of the ATSC Transport Stream

known as *multicasting*. (It is in some contexts important to distinguish ATSC multicasts from the analogous HD Radio multicasts described in Chapter 13, or from the quite different *IP multicast* format used on the Internet and in AoIP or Video-over-IP systems, described in Chapters 6, 8 and 9.)

Typically, broadcasters wanting to maintain the highest video quality do not exceed one HD program, or four SD programs in the multiplex at one time. It is quite common, however, to multicast one HD and one SD program, or add even more SD channels if they include mainly static pictures, such as graphics or weather radar. It is theoretically possible for a station to change its program mix at different times of the day, perhaps broadcasting four SD programs during the day, and switching to one HD and one SD program in the evening. This rarely happens in practice, however, because stations have found it preferable for viewers and cable companies if their channel lineup remains constant.

It is important that the total data rate for all of the program streams added together never exceeds the total payload data rate available in the DTV channel (19.39 Mbps). For example, a DTV channel might be set to multicast four different programs at various bit rates that could work well most of the time. But if all four programs had content with a lot of motion, highly detailed images and/ or rapid scene changes at the same time, the single DTV channel might not be able to handle it, and visible compression artifacts occurring on one or more of the services could result during those passages.

Statistical Multiplexing

A technique that can be used to help maintain the highest video quality while multicasting is called *statistical multiplexing* (sometimes referred to as a "statmux" or statmuxing"). This takes advantage of the fact that on average (i.e., statistically), different programs have different amounts of picture detail and motion at any one time. The multiplexer varies the bit rate allocated to each program on demand, to allow each one the bit rate it needs to maintain high quality at that moment. It will be rare for all programs to require maximum bits at the same time, which could result in *bit starvation*, causing compression artifacts.

CLOSED CAPTIONS

In NTSC, the information needed for carrying closed captions for the hearing impaired is carried as data on line 21 in the vertical blanking interval. In ATSC the captioning data is carried in another way. Within the video compression system, space is reserved in an area called the *video user bits* for carrying this additional data.

The DTV closed captioning system is specified in the Consumer Electronics Association (CEA) standard CEA-708-E. When this system was designed, the opportunity was taken to enhance the capabilities compared to NTSC captions. Thus, DTV captions can have a choice of fonts in different colors and sizes and at different locations on the screen, several caption services can be provided simultaneously (for example, adding captions for different languages or for "easy reading") and there are other enhancements available.

PROGRAM AND SYSTEM INFORMATION PROTOCOL (PSIP)

FCC rules require broadcasters to carry *Program and System Information Protocol* information, usually known as "PSIP," in accordance with ATSC standard A/65. This is transmitted in the multiplex with the along with the DTV program(s) bitstream. PSIP helps the viewer navigate to the correct program, provides an electronic program guide for current and future programs, and helps the DTV receiver tune to the correct channel and correctly decode the program.

PSIP is carried as a series of data tables that are regularly updated and carried in the DTV "pipe," as shown conceptually in Figure 16.9 above.

Major and Minor Channels

As explained in the section on Channel Navigation and Branding earlier in this chapter, PSIP allows broadcasters' DTV channels to be identified with the channel number that relates to their earlier NTSC channel number, whether or not they are actually still using that channel for their broadcasts.

PSIP includes a *Virtual Channel Table* (VCT) that, for terrestrial broadcasting, defines each DTV service with a two-part number consisting of a *major channel* followed by a *minor channel*. The major channel number is the one mentioned above (usually the same as the NTSC channel previously assigned to the station), and the minor channels have numbers depending on how many DTV services are present in the DTV multiplex, usually starting at 1. Nomenclature usually separates the two numbers with a hyphen (5-1, 5-2, etc.), although sometimes a decimal point is used (9.1, 9.2, etc.). Different arrangements, using single numbers, are used for ATSC services carried on cable channels. The VCT also includes the PSIP *Short Name* given to each service. A station might label one service as their call letters with the suffix "HD" and, if they are multicasting, another might be labeled as SD, or some other name.

As an example, a public television station in Washington, D.C., WETA-TV, which broadcasts its DTV service on UHF channel 27, had originally broadcast its NTSC service on UHF channel 26, and at this writing identifies its four over-the-air digital multicast services via PSIP as shown in the two left columns of Table 16.2 below. (The rightmost column is presented for informational purposes here.)

Services consisting only of data, with no video or audio, must use minor channel number 100 or greater.

Electronic Program Guide (EPG)

PSIP includes *Event Information Tables* (EITs) that provide a program schedule for each DTV program service, which allows the receiver to build and display an electronic program guide (EPG). The EITs should carry information for at least the next 9 to 12 hours of programs, and may include information up to a maximum of 16 days ahead. As well as the basic data for each program of title and start time, the information should include *content advisory* rating, audio format (for example, stereo or 5.1), language, and closed caption details. Extended Text Tables (ETTs) may carry additional descriptive information about each program that may be displayed.

TABLE 16.2
Example of a DTV Station's Multicast Offerings

Channel	*PSIP Short Name*	*Content*
26-1	WETA-HD	Primary content, High-definition (720p)
26-2	CREATE	Alternate content 1, Standard-definition (480i)
26-3	KIDS	Alternate content 2, Standard-definition (480i)
26-4	TV26	Primary content, Standard-definition (480i)

Directed Channel Change

Directed Channel Change (DCC) is an interesting feature that allows broadcasters to tailor programming or advertising based on viewer demographics. For example, viewers who enter location information, such as their zip code, into a DCC-equipped receiver could receive commercials that provide specific information about retail stores in their neighborhood. Segments of newscasts, such as weather alerts that are relevant to certain areas, could also be targeted based on this location information. A channel change may also be based on the program rating or the subject matter of the content of the program. Nearly 140 categories of subject matter have been defined that can be assigned to describe the content of a program. A broadcaster can use this category of DCC request switching to direct a viewer to a program based on the viewer's desire to receive content of that subject matter.

DATA BROADCASTING AND INTERACTIVE TELEVISION

As previously stated, the transmission system for digital television can be thought of conceptually as a large pipe for carrying the digital bitstream, which is divided up into smaller pipes for different program services, each of which carries bits of information representing a content type. It is perhaps not surprising that, in addition to regular television programs, the DTV pipe can be used to carry bits representing other types of digital information, some of which may not have any relation to television content. This is generally referred to as *data broadcasting* (also known as *datacasting*).

Bandwidth

As discussed previously, the amount of bits within the 19.39 Mbps available in the ATSC channel's payload that are needed for television programs' video and audio depends on the number and type of programs carried. It is possible that there may be spare bandwidth that is not needed for video and audio programs, and this may vary considerably at different times of day or even from instant to instant, depending on picture content.

A particular television station decides how much of its bandwidth to allocate to data broadcast services (if any); this may be a fixed amount ranging from a few kilobits per second to several megabits per second, the latter being comparable to a reasonably high-speed broadband Internet connection's downlink.

Opportunistic Data

Even if no fixed bandwidth is allocated for datacasting, there will always be some spare bits available in the bitstream due to the variations in picture content in the television programs. These spare bits can be used for *opportunistic data* services that are delivered at an indeterminate data rate. This data is transmitted whenever an opportunity becomes available in the ATSC signal. For example, even if a high definition program is being transmitted, there will be times during the program when the data rate necessary to carry the video information will be reduced, such as when a still picture is on the screen. At that point, more capacity is available for opportunistic data services. Such service is most appropriate for applications in which latency (i.e., delivery delay time) is of little concern.

Data Piping, Data Streaming, and Data Carousel

Depending on the type of datacasting service, the data required may be sent as a "package delivery" (for example, downloading a file) or as "streaming" for continuous data delivery.

In some cases, the data may be sent repeatedly in a *data carousel*, which provides more opportunities for downloading the data to the receiver.

Types of Data Services

There are two basic types of datacasting service: those that are associated with a regular television program, known as *program-related data*, and those that standalone, or *non-program-related data*. Examples of program-related services include the following:

- Supplementary information about the content of the program ("tell me more" data)
- Sports statistics and additional game or player information
- Opinion polls or voting as part of a program
- E-commerce applications for products or services featured in a program

Some DTV receivers may be designed to display the additional program-related data services, inserted over the program pictures, with picture-in-picture techniques, or with separate screens, all selectable on the remote control.

Examples of non-program-related services include the following:

- News, weather, stock quotes
- Web pages
- Newspaper/magazine downloads
- Music/movies downloads
- Software downloads

Some DTV receivers may be designed to display the additional non-program-related data services in some manner, selectable on the remote control. In other cases the data may be intended for use on a computer or some other device, and will not run directly on a television.

Interactive Services

Interactive service implies that the viewer can respond to the data service, sending messages to request more information or services, or to provide feedback. This often requires the use of a *back-channel* (also known as a *return-channel*) that sends the viewer's response or request back to the source of the service at the broadcast station or elsewhere. Digital cable television systems have the capability for a back-channel from the home built into the system, because the cable can transport information in both directions. Because over-the-air broadcasting is one-way only, however, broadcasters have to provide an alternative path for the return signals. Originally, dial-up telephone lines were used for this purpose (and still may be in some cases), but today, the viewer's existing Internet connection is more commonly employed. (Dedicated wireless return paths for interactive broadcast back-channels have been proposed, but never successfully deployed.)

Examples of true interactive services include quiz or game show participation, polling, e-commerce, and any service that requires additional information to be sent from the source of the data service upon request from the viewer. There are some services that appear to be interactive that do not actually require a back-channel, however. In such cases, all of the data that relates to a particular feature (for example, supplementary program information) is downloaded automatically to the receiver via the broadcast, but is not immediately displayed. Instead it is stored in the receiver, and remains ready for display if and when the viewer requests it.

Data Broadcast Receivers

Some DTV receivers and set-top boxes are designed to display the additional program-related data services. Typically, these receivers have additional buttons on their remote controls for the request and display of this additional information, and the receiver has memory to store at least some of the data received. Other types of data services are intended to be used on a computer. Computers can have a DTV tuner added to receive such services or, in some cases, it may be possible to connect a computer to a DTV set-top box. With the convergence of technologies, the line between television and computer devices is becoming blurred, and television receivers with hard disk storage and an Internet connection (often referred to as "smart TVs" or "connected TVs") are increasingly common. Eventually it is expected that most television-viewing devices will have network capabilities, allowing the possibility of accessing, sharing, and displaying all forms of digital media across all of a user's media devices. (See Chapters 17 and 18 for more on this topic.)

CONTENT PROTECTION (CONDITIONAL ACCESS)

The ATSC system offers the ability to restrict a user's ability to view certain content via the addition of *Conditional Access* (CA) control. This allows broadcasters to charge a fee—or at least require a user to register—for access to a particular service, either on a subscription or a pay-per-use basis. Use of this approach involves encryption of the protected content by the broadcaster, and the delivery of a key (and possibly other software) to authorized users to enable their decryption of the content in their receivers. Arrangements for this are similar to those used by cable and satellite companies for premium TV channels and pay-per-view, and usually require a smart card or other authentication method to be used for access, with charges billed to a user's credit card, banking, or other account. CA can be applied to regular television content or to datacasts.

ADVANCED ATSC SERVICES

Subsequent to the deployment of the initial ATSC system and the U.S. DTV transition, several extensions to the system have been developed. Each of these services has been developed in a *backward compatible* manner, meaning that a broadcaster's addition of any such service will be ignored by existing ("legacy") receivers, and not affect the experience of those users, while users with new or upgraded devices will be able to access both the original and the extended services.

Mobile DTV

ATSC DTV was developed to be received on fixed receivers, generally with external ("rooftop") antennas. This seemed appropriate in the period of the standard's original design (mid 1990s), but subsequently a desire to deliver broadcast TV content to mobile devices has emerged. To address this, ATSC developed a backward-compatible, add-on capability for *Mobile DTV*, the standard for which was initially published in 2009 (ATSC A/153). ATSC Mobile DTV is sometimes referred to as *ATSC M/H*, (for Mobile/Handheld).

The service is multiplexed into the ATSC channel as an A/V (or audio-only) streaming-media data broadcast, with an identity that is ignored by fixed receivers. The Mobile DTV service includes a large amount of additional error correction so that it can be more robustly received than traditional ATSC service. It also uses more recently developed codecs (MPEG-4 AVC for video, and MPEG-4 HE-AACv2 for audio), as well as other technologies not available at the time of the original ATSC format's development. A mobile emergency alerting component (M-EAS) was added in 2013. A variety of handheld and mobile receivers have been developed, along with hardware adapters ("dongles") that add Mobile DTV reception capability via plugging into USB ports on existing computers and mobile devices.

Non-Real-Time Broadcasting

Another addition to the ATSC system permits non-real-time (NRT) content delivery, which allows a broadcaster to deliver file-based content to users for on-demand viewing, without requirement for the use of a backchannel. The service is defined in ATSC A/103. It offers three different "consumption models": *Browse and Download*, *Push*, and *Portal*.

In the Browse and Download model, the user is presented with a catalog of available content, which can be ordered for subsequent delivery via over the air datacast. "Ordering" in this sense means that the receiver is alerted to watch for data packets associated with the selected program, and when they are detected in the broadcast, to capture and store them, then notify the user that the program is available for playback.

Push mode allows the user to subscribe to a particular service offered by the broadcaster, which can include text, graphics or short media clips that are downloaded and continually updated, for immediate display when the user requests them.

The Portal model broadcasts files that when downloaded act like a webpage or website, which the user can call to the TV screen when desired. Portal files can serve as a home page for the station or a particular TV program, for example.

3D TV

At this writing ATSC is preparing to publish a standard for adding 3D TV capability to ATSC broadcasts, which would provide a variety of methods for broadcasters to compatibly offer 3D content to users.

ATSC 2.0

Another pending development at the time of this writing is a suite of backward-compatible ATSC extensions collectively referred to as *ATSC 2.0*, which address the emerging connected television environment. ATSC 2.0 is a "toolbox" of components that offer capabilities such as interactivity, second screen services, hybrid television, additional content security, new coding algorithms, additional NRT features, enhanced electronic program guide, and measurement systems for reporting audience usage of these features. Some or all of these capabilities could be added to ATSC broadcasts and receivers over time.

CHAPTER 17

Alternate Television Delivery Systems

Like radio, the television industry has also been deeply affected by the coming of the Internet and the alternate content-delivery methods that it affords broadcasters and others. But there are also significant differences in the way the technology has been applied across radio and television, which are largely functions of business, policy or regulatory issues rather than technical distinctions between the industries. This chapter will address both the technical and the non-technical axes of this emerging technology and its applications in television.

INTERNET TELEVISION STREAMING AND DOWNLOADING

Again as in radio, perceptual coding algorithms and streaming media encoding (see Chapter 14) are applied to video for its distribution via the Internet. Television content is thereby available online for on-demand streaming as well as via file downloads for later viewing, although the data rates required are at least an order of magnitude higher than those required for radio.

Another important difference between online radio and TV service is the quality levels that must be supported. A radio station typically offers its streaming service at a single data-compression level and bit rate, or at most it may offer two different levels (perhaps using different codecs on each), with relatively minor differences between them. While lower bit rates, if they are offered, are often preferred by wireless users to reduce their data consumption costs, audio streams of any data rate can typically be listened to on any streaming device with acceptable results. This is decidedly not the case for video, where higher data-compression levels and lower bit rates may be acceptable on the small displays of handheld devices and minimized media player windows on PCs, but much higher data rates must be provided for acceptable full-screen viewing on televisions—and even high-resolution tablet displays—particularly for high-value content like movies, episodic television and sports.

This additional bandwidth requirement has made a hybrid of streaming and downloading called *progressive download* to become quite popular in online video, by which a file can be requested for download from a website, but it can begin playing on the user's device prior to completion of the download. After the file download begins, the device's media player for the service gauges how quickly the download is progressing, and allows the file to begin playing at such time as the player estimates the download will always remain ahead of the playback. Should the device's Internet connectivity suffer a change that substantially slows or temporarily stops the download rate after playback has started, however, the playback may "catch up" to the full extent of the download and the player will run out of cached material, causing the playback to be interrupted until the download resumes and an adequate cache is restored. Media players designed for progressive download usually include in their user interface a two-level progress bar that shows both the amount of the file that has been downloaded and the playback progress of the file.

Scaling Techniques

The relatively high data rates required for high-quality online video streaming are challenging to provide. Even with state-of-the-art data compression, multi-megabit connectivity is usually required, especially for streaming-media quality suitable for full-screen presentation on large-screen televisions. Ever-growing broadband Internet capacity and speeds have made this possible since the mid to late 2000s, however, when fixed broadband connections to U.S. homes began to reach or exceed the 1 Mbps mark.

Yet because most video streaming takes place largely over unicast IP connections (as explained in Chapters 6 and 14), even if each user's "last mile" connectivity could support the requisite speeds, the Internet backbone would be taxed to supply many users' simultaneous requests for streaming video. To accommodate this in ways beyond simple brute-force upgrading of backbone capacity, the Content Delivery Network (CDN) approach explained in Chapter 14 provides substantial assistance. It allows delivery of streaming media content from numerous, decentralized and *mirrored* servers, each storing the same content in different physical locations distributed across the Internet. This allows users in different geographic areas to select content that is delivered to them from servers that are closest to them along the edges of the network, rather than all users' requests being concentrated to a single location. This avoids both overtaxing of the servers at that one location, as well as alleviating congestion of the Internet backbone in proximity to a single, central server location.

Another technology that attempts to address this issue is *adaptive streaming*, by which video content can be encoded at several bit rates and presented for streaming as a single offering. The viewing device automatically selects the highest quality version that current network conditions will allow. In its latest incarnation at this writing, *dynamic adaptive streaming* performs such encoding of a program by breaking it into many smaller files and encoding each segment at the various bit rates. The program is presented to the viewer as a kind of progressive download, with the viewing device continually monitoring current network conditions and deciding as the end of each segment approaches whether to stay at the current encoding level, or to up- or down-shift to a different encoding level, based on whether the device estimates that the next segment will download in time to be viewed without interruption given the connection's current status. A number of companies have developed proprietary methods of providing this capability, and a standard has been developed by MPEG called *Dynamic Adaptive Streaming over HTTP* (DASH), often referred to simply as *MPEG-DASH*.

TV Broadcast Services vs. Online Content

Chapter 14 explained that many radio stations routinely stream their over-the-air content services online on a full-time basis. This rarely occurs in local television, however, due to the higher reliance of television stations on content acquired from outside entities, rather than being produced by the station itself. Most of this externally produced content is licensed to the station for over-the-air broadcast only, and with distribution rights (usually exclusive) for the station's local service area only. Because the Internet is largely a global service, and because it is generally considered to be a different delivery method in licensing agreements, most television stations are not able to stream their over-the-air content services online, except when they are airing content originating from their own production studios (such as local news programs), or other content for which they own full distribution rights. Meanwhile, the original content providers (television networks, syndicators, etc.) often retain the rights to stream their content online themselves, setting up a potentially competitive relationship with their affiliates across multiple platforms. As a result, such direct online offerings by the content owner are also a subject of negotiation in affiliation agreements.

On occasion, externally acquired content also may not be able to be streamed by <u>radio</u> stations (e.g., certain syndicated talk-radio programs), but it is the exception rather than the rule in that industry, whereas the opposite case exists for television.

This creates an environment in which the vast majority of online television usage is program-based rather than service-specific. Thus Internet television users generally search for individual <u>programs</u>, as opposed to seeking "channels" as they often do in Internet radio. Such a usage model includes episodic television programs, which typically become available online at some short period of time (e.g., one day or one week) after the program is broadcast on air.

Because online episodes often remain available as new episodes appear, viewers can watch all previously aired episodes of a series, a process sometimes referred to as *"Catch-up TV."* In some cases, a complete series of (typically non-broadcast) episodic TV programs is released simultaneously for online access. The availability of large numbers of television-series episodes simultaneously online has led to a consumer behavior known as *binge viewing*, by which a user watches several episodes of the program (or even the entire series) back-to-back in a single session, or over a short time period.

Meanwhile, television stations' websites often offer the stations' own locally produced content for on-demand streaming of their latest newscasts, weather forecasts or traffic reports, for example.

This implies that viewers of local television stations are limited to receiving most if not all of these stations' content in real time only from traditional sources at this writing (i.e., over the air, or via cable, satellite, and IPTV / Telco TV services), as opposed to local radio station listeners, who typically have their choice of listening to local stations over the air or online—although perhaps at additional cost for the latter. Note that there have been attempts by a number of third parties to offer real-time streams of broadcast television stations' content via the Internet, but these have all generated litigation from various parties questioning their legality, much of which remains unsettled at this writing, so it is unclear whether such service will ever be legally offered to users. Various technical and legal / regulatory solutions have been proposed as solutions to this important issue, but none has yet emerged as a broadlyacceptable method for the provision of such service as of this writing.

Aggregation Sites ("Over-the-Top" TV)

This has led to the popularity of online video content aggregation sites that are centered on listings of television and cinematic programs, rather than simply linking to broadcast station streams, as such sites typically do for Internet radio. Most of these video aggregation sites also include rich search, review, and recommendation features, and a few have become quite popular, with research indicating that a large concentration of Internet traffic is devoted to video streaming from these sites during evening prime-time viewing hours. Most of this is directed to the professionally authored content available from movie studios and broadcasters, but a significant amount of viewing also arises from *user-generated content* (UGC) sites, which are open for posting of video material from anyone. For example, at this writing, it is estimated that approximately half of North American downstream Internet traffic during peak usage hours emanates from only two of the most popular online video services—one a movie and TV-show site (called "Netflix") and the other a primarily UGC site (called "YouTube").

Some of these services are provided free, while others use a subscription or a pay-per-view model. The latter are typically provided without commercials. Because these services use the viewer's existing Internet connection to deliver their service (as opposed to providing their own dedicated connectivity like a broadcast transmitter, cable network, or satellite system), they are called *Over-the-Top* (OTT) services. The OTT video distribution model is largely replacing the traditional video rental business (using physical media), although DVD and Blu-ray disk rentals still persist at this writing, largely via kiosks. Meanwhile, traditional cable, DBS and IPTV/Telco television services' *video on demand* (VOD) features, which offer similar capabilities but over their own networks, are also being impacted by OTT, and some have altered their VOD services to appear more "OTT-like" in their offerings and user interfaces.

OTT services offering professionally produced video are operated by a variety of entities, including some broadcast networks, and they offer a wide range of movies and television programs. Local station content is not typically offered by these sites, although some stations routinely post their content to one or more UGC-focused OTT sites.

CONNECTED TELEVISION

Connected Television is a label used to describe the growing trend among televisions to include Internet connectivity as a source of content. This is either provided via a wired Ethernet connector on the TV, or via internal WiFi connectivity, or both. At this writing, it is difficult to find a current model large-screen television without such capability built in, although not all purchasers of such sets are believed to be actually connecting them to the Internet. Nevertheless, this has led to the growth of the online television viewing movement described above.

The Connected TV movement also applies to older televisions that do not have such connectivity built in, because they can be connected to a variety of outboard television devices that provide Internet video access, such as video game consoles, Blu-ray players, or dedicated online TV set-top boxes. In addition, a class of devices called *HDMI sticks* are small units akin to USB flash drives that plug into an HDMI port on digital televisions, and provide WiFi connectivity to the user's home network. Most of these devices allow the user to share content from other devices on the home network, for display on the main digital television screen. They also offer connectivity via the user's home network and broadband connection to the Internet, where they can be used to browse the web, and/or to visit specific OTT sites offering on-demand video content, via applications downloaded or natively installed on the device. (In some cases, a particular television or external hardware device may be limited to accessing only certain providers' sites for such content.)

VIDEO PODCASTING

Just as in radio, video podcasting allows a content provider to offer frequently refreshed content to users on a subscription basis via the Internet, received via a podcasting client on the user's device(s). This allows local television stations (and other video content creators) to present viewers with their content via automatic download, so that the latest version is available on the user's device, for immediate viewing access whenever the viewer requests it—even if the user's device is offline at the time.

MOBILE TELEVISION "APPS"

As broadband <u>wireless</u> Internet access availability and speeds have increased, mobile video usage has grown dramatically. From nearly non-existent prior to the late 2000s, mobile video has become one of the primary applications on smartphones and tablets at this writing. A large number of mobile "apps" are available to enable easy access to such video content from various providers. These apps typically include one or more audio and video codecs and a custom user interface to facilitate access and viewing controls.

Streaming video is also available to mobile devices from the web, so the browser on the mobile device can be used for access, in lieu of a specific mobile video app.

Because display screens on mobile devices are small, and wireless broadband connectivity is generally more constrained than fixed Internet access, video content tailored for mobile access is often offered at lower resolutions, and therefore requires lower data rates for streaming or downloading.

SECOND SCREEN AND SOCIAL TV

Research has shown strong evidence of consumer behavior involving the use of online devices while watching television. Because some of this activity is related to the TV programs currently being watched, content creators, broadcasters and third parties have developed applications for online devices designed to be used in conjunction with TV programs. These applications are generally referred to as *Second Screen* apps, and they typically deliver program-associated content, enhancement data, quizzes and polling, or other content related to the TV program via the Internet connection of the Second Screen device. In some cases, the Second Screen content is dynamic, and changes as the program progresses. Second Screen is sometimes called *Supplemental Screen*, since there may be more than two in simultaneous use by the television viewer, and sometimes the supplemental screen becomes the primary focus of attention. Finally, in some cases, the primary video content may be shifted to the user's portable device, for continued viewing while the user is in transit between locations having larger, fixed screens.

Second Screen applications can also be used for viewer feedback in response to the program, including e-commerce activities via the device's Internet connection, for purchase of items shown during the program or a commercial announcement, for example.

Automatic Content Recognition

There are numerous ways in which Second Screen content is located by the online device—and if necessary, appropriately synchronized with the corresponding segment of the television program currently being viewed by the user. For live programming (such as sports), the Second-Screen data is simply distributed via the Internet in as close to real time as possible. But for the majority of programs, different viewers may watch the program at different times due to time zone variations, the scheduling preferences of their local stations, or playback from their own DVRs. To accommodate such differences in viewing times, a robust and versatile method of calling up the appropriate application on the Second Screen device, and maintaining its content in synchronization with the program's progress, is called *Automatic Content Recognition* (ACR).

The most common form of ACR at this writing uses the microphone in the Second Screen device to acoustically pick up the soundtrack of the television program emanating from loudspeakers in the viewing location. An app on the Second Screen device listens to this audio and uses it as a reference to locate and display the appropriate online content that accompanies the television program.

In some cases, the app uses the waveform of the television program's audio, which it attempts to match to a copy of the waveform stored in an online database at a site to which the Second Screen app connects—a technique known as *fingerprinting*. Alternatively, a small amount of data is inaudibly embedded in the soundtrack of the television program, and repeated frequently. In this case, when the microphone of the Second Screen device picks up the soundtrack, the Second Screen app detects the embedded data and uses it to identify and navigate to the appropriate content; this technique is called *watermarking*.

Social TV

A related application of Second Screen usage is conversations among viewers of the same TV program via social media. These conversations or postings can be conducted on popular, generic social media sites, or on one of many television-specific "Social TV" sites, where all the postings on a given page, sub-site or thread are related to a specific TV program or channel, and can be filtered to the user's preferences.

Note that with all Social TV and Second Screen applications, multiple individual users all watching the same program together on their shared primary screen can be engaged in separate, individual experiences on their respective Second Screens. Such "collective personalization" is a key feature of the Second Screen movement.

HYBRID TV

The discussion of Connected TV above assumes that televisions with the Internet connectivity use <u>either</u> a traditional delivery service (e.g., over-the-air broadcast, cable, DBS, or IPTV/Telco TV) <u>or</u> the Internet as an independent source for content, with users switching between these like they would change channels, or switch from a DVR to a Blu-ray player, for example. Another option is the use of a traditional source <u>and</u> the Internet as simultaneous sources of content, however, where some elements of a single viewing experience come from the traditional source, while others come via the Internet simultaneously, without the user really knowing or caring which elements came from what source. This so-called *Hybrid TV* approach allows personalization of a broadcast program, in which the common content to all viewers comes via the traditional broadcast source, while the elements that can be custom-selected for display by a particular user are supplied via the Internet. This additional content can be displayed on the primary screen or on a supplemental device. An example is the selective display of additional statistics overlaying the video coverage of a sporting event.

In some cases, the Internet-delivered content may be a video segment that replaces a segment of content in the main broadcast feed in a synchronized fashion. An example of this is *targeted advertising*, where an ad intended for a specific viewer or group of viewers is delivered via the Internet, replacing the generally delivered (untargeted) ad or other announcement in the main broadcast.

Technical standards have been developed in various regions of the world for Hybrid TV, and are expected at this writing to continue expanding their capabilities over time.

CHAPTER 18

Next-Generation Broadcast Television Systems

In Chapter 16, the ATSC Digital Television is described in its current form at this writing. That description included the original ATSC DTV system (now retrospectively called "ATSC 1.0"), as well as some of its backwardly compatible upgrades known as ATSC 2.0. This chapter will summarize what is envisioned for the next generation of ATSC DTV, called *ATSC 3.0*.

As mentioned in Chapter 16, the ATSC DTV format is used for terrestrial digital television broadcasting in North America, S. Korea, and parts of Central America. Other parts of the world use one of the three other worldwide DTV format "families" detailed in Chapter 2—DVB, ISDB, or DTMB—or are still broadcasting analog television and have yet to convert to any of these standards (although decisions may already have been made on which format they will convert to in the future, and when). Each of these other formats is also considering improvements in their respective services, and some have already been deployed. Later in this chapter some specifics of these international developments will be discussed, but by and large they are all considering similar kinds of improvements, which will be generally described next.

PROPOSED DIFFERENCES FROM CURRENT SYSTEMS

There are a number of consumer trends and changing preferences that next-generation television systems are attempting to address. Most of these involve the disruptive influences arising from the convergence of broadcast and online services, but others are simply enabled by the continuing evolution of consumer electronics technologies propelled by Moore's Law. Among the latter is a pursuit of ever-greater sensations of realism in the media consumption experience, which several key elements of next-generation television proposals attempt to address. The following analysis will focus primarily on ATSC 3.0 as an example of all such developments.

ATSC 3.0 Usage Scenarios

To maximize the value of its next-generation system, ATSC decided early in its development process to absolve the new system from the necessity of full backward compatibility. Thus unlike ATSC 2.0, the ATSC 3.0 format may or may not be compatible to all elements of the predecessor ATSC system. This allows the next-generation system to incorporate new codecs, new modulation schemes and a wide variety of other categorical (rather than simply incremental) improvements. But if a non-compatible approach is taken, the improvements of the new system must be adequately substantial in order to warrant the upheaval of broadcasters' and consumers' switching to new equipment, in what would amount to a second digital television transition.

ATSC identified 13 areas of expanded capability that the new system might address, as follows:

- **Flexible Use of Spectrum:** The ability to use "smart" or more adaptive methods of allocating bandwidth, to enable more efficient use of broadcast spectrum.
- **Robustness:** A high degree of resilience to noise, attenuation and interference, allowing reliable off-air reception for a wide variety of devices and locations.
- **Mobile:** Strong emphasis on mobile television reception on handheld and vehicular receivers, with up to HD video quality.
- **Ultra HD:** The ability to broadcast Ultra High Definition (UHD) video content of at least 3840×2160 resolution, and possibly with frames rates greater than 60 Hz, increased luminance dynamic range, improved color resolution and expanded color gamut (see Chapter 5).
- **Hybrid Services:** Intrinsic capability to blend over-the-air and online content-delivery modes for a rich, bidirectional multimedia consumer experience.
- **Multi-view/Multi-screen:** Capacity to transmit and display multiple independent content elements simultaneously on one or more user screens.
- **3D Video:** Well-integrated 3D video capability (see Chapter 5).
- **Enhanced and Immersive Audio:** Taking TV sound to the next level, with full spatial reproduction and added realism, plus multiple options for user adaptation and adjustment.
- **Accessibility:** Integrated features for visually and aurally impaired users, plus multiple language support, and dialog enhancement capability.
- **Advanced Emergency Alerting:** Improved capabilities for targeted, localized alerting, including for visually or aurally impaired users, and activation when receivers are off.

- **Personalization/Interactivity:** The ability for individual users to tailor a mass-distributed medium to their own preferences, and to directly respond to the content.
- **Advanced Advertising/Monetization:** Offering new models of support via targeted advertising and other commercial techniques not previously possible with a traditional broadcast service.
- **Common World Standard:** Providing economies of scale through the use of unified technologies across all next-generation world DTV formats (see below).

The attributes above are indicative of what all DTV formats are considering as successors to their current standards. They address ways in which television can adapt to new user behaviors and preferences, as well as offering broadcasters new methods of connecting with audiences. Note that some of these attributes are difficult to fulfill within a single methodology because their requirements are mutually exclusive. For example, providing both high data-rate delivery for UHD and highly robust service for ubiquitous mobile reception are challenging to provide optimally in a single transmission mode. Therefore a key element of next-generation system design is the ability to provide a wide range of simultaneous, independent services at the broadcaster's option within a single-service multiplex, such as a lower-resolution robust service targeting mobile reception and a less robust but higher quality service for fixed reception, both contained within the same RF transmission channel.

They can be summarized further into three main categories: (1) Robust, ubiquitously available delivery platform; (2) Rich and immersive user experience; and (3) Combining one-to-one, one-to-many and many-to-many architectures for a flexible and open-ended worldwide media ecosystem.

Other Key Influences

Several technologies are likely to be key factors in the next generation of broadcast television. These include advanced applications of the Orthogonal Frequency Division Multiplexing (OFDM) technology that has been used in some current-generation DTV systems (although not in ATSC 1.0), along with the latest ISO/MPEG video codec, *High Efficiency Video Coding* (HEVC, also known as ITU-R H.265), which provides about twice the efficiency of MPEG-4 AVC (ITU-R H.264—see Chapter 16), and four times the efficiency of the MPEG-2 video coding used by ATSC 1.0. It is also likely that next-generation broadcast systems will gravitate toward Internet Protocol (IP) as a core transport technology, moving away from dedicated TV-centric transport approaches used in current systems.

It should be recognized that broadband wireless systems have developed broadcast-style methodologies in their latest generation (4G) technology, called *Long Term Evolution* (LTE) *Advanced*, as specified by the *3GPP* organization, which includes a mode entitled *Multimedia Broadcast/Multicast Service* (MBMS). LTE also applies OFDM in a cellular transmission architecture, and its MBMS mode could emerge as either competitive or complementary to next-generation broadcast systems.

Another guiding element to next-generation TV will be the movement to higher quality video and audio. The *International Telecommunication Union Radiocommunication Sector* (ITU-R) has produced its *Recommendation BT.2020* on UHD, which has served as foundation for this technology's progression into future broadcast delivery systems. Meanwhile the Society of Motion Picture and Television Engineers (SMPTE) has produced similar guidance for the content creation community on UHD in its *SMPTE 2036* document, and the Consumer Electronics Association (CEA) is also working on UHD standards for consumer systems. More efforts are expected from each of these groups, and others, which will further inform the development of next-generation TV delivery systems over time.

An overview of related developments worldwide at this writing is presented in the following section.

SYSTEM PROPOSALS AROUND THE WORLD

DVB

As explained in Chapter 2, the Digital Video Broadcasting (DVB) family of DTV formats is used throughout Europe and much of Asia and Africa, in Australia and New Zealand, as well as in some Latin American countries. In addition to terrestrial television broadcasting (DVB-T), it also publishes standards for cable television (DVB-C) and satellite television (DVB-S), each of which is also widely used in various locations around the world. DVB has in effect already moved to a set of next-generation standards with its DVB-T2, DVB-C2 and DVB-S2 standards, all initially published in the mid-to-late 2000s by the *European Telecommunications Standards Institute* (ETSI). Note, however, that the first-generation DTV broadcast system specified by DVB-T did <u>not</u> include provisions for high-definition television, as its contemporary ATSC A/53 ("ATSC 1.0") did. So it can be argued that DVB-T2 puts DVB at parity with ATSC A/53 for services it can deliver. But owing to its development about a decade after A/53, DVB-T2 takes advantage of some significant transmission advancements in the interim, and therefore stands as a more advanced delivery system overall at this writing. (ATSC 3.0 is considering many of these same improvements and those that have continued to advance in the interim, to meet or exceed DVB-T2's benchmarks for DTV delivery.)

DVB-T2 has been deployed in a few countries at this writing, and meanwhile DVB continues to develop additional standards that can be considered truly next-generation TV, particularly the DVB-NGH (Next-Generation Handheld) standard, targeting the same robust delivery noted above in ATSC 3.0's target attributes.

ISDB

Integrated Service Digital Broadcasting's terrestrial DTV format (ISDB-T) is used in Japan, the Philippines, and most of South and Central America, as detailed in Chapter 2. The Association of Radio Industries and Broadcasters (ARIB), the Japan-based organization that developed ISDB-T, is also considering next-generation projects. Chief among them is the *Super Hi-Vision* format under development by Japanese national broadcaster NHK. It defines a UHD-2 format (7680 × 4320 pixels, with frame rates up to 120 Hz), so it currently stands as the system requiring the highest data capacity proposed for next-generation DTV delivery. At this writing, it is targeted for terrestrial deployment in Japan around 2020, with DBS delivery possibly sooner.

DTMB

Digital Terrestrial Multimedia Broadcasting (DTMB) is a system developed under the auspices of China's *State Administration of Radio, Film and Television* (SARFT), which is currently used in China, and expected to be deployed in Cuba at this writing. Meanwhile SARFT is developing a Next Generation Broadcasting—Wireless (NGB-W) system, which anticipates many of the same capabilities as noted above under ATSC 3.0, with the addition of a dedicated backchannel system for TV viewers to use as a return path for interactive features. This backchannel is primarily intended for use in rural areas of China, where other forms of bidirectional data communication (such as wireless Internet service) is not commonly available.

FOBTV

An organization called *Future of Broadcast Television* (FOBTV) was formed in 2011 by a group of organizations and companies including all of the above DTV standards bodies and others, with the intent of harmonization of their next-generation systems. The organization's goal is not to create additional standards but to provide a forum for coordination and possible alignment of international DTV standards, in whole or in part, such that the next generation of DTV systems will be less divergent than current systems.

NEW DIRECTIONS IN AUDIENCE MEASUREMENT

As with any media service, an important part of the broadcast business is accurate measurement of its audiences. Just as the media industry itself has been disrupted by digital technologies, so too have the methods of its audience assessments. In its earliest incarnation, audience research was conducted by polling selected samples of the audience via handwritten diaries sent through the mail, by which these consumers reported their broadcast-service usage over a given time period. This manual method has been gradually replaced in recent years with more high-tech systems that automatically record audience behavior, and as a result provide far more reliable data on audience viewing and listening. Challenges remain, however, regarding fair and accurate reporting of time-shifted consumption of broadcast content, and the aggregation of on-air and online usage of the same content.

It is therefore no surprise that developers of next-generation broadcast systems are exploring methods of integrating technologies that enable accurate audience measurement of all services provided. At this writing, advanced watermarking systems appear to be a key element of this trend, allowing imperceptibly embedded data to be added to broadcast content, with unique watermarks for each channel and each delivery platform. Initial implementations of such watermarking are already in use for audience research purposes, as explained in Chapter 8. This provides greater statistical accuracy in the estimation of the overall audience for a particular content segment, as well as being able to determine the method by which it was delivered to each sampled user.

Another important difference between broadcast and online services is that while broadcast delivery must be approximated by surveys and statistical analysis, online delivery can be directly and accurately measured owing to its bidirectional connectivity. This presents another challenge to audience measurement that attempts to fairly accommodate both paradigms. Moreover, as future broadcast systems envision a move toward a majority of receivers that include online capabilities, direct audience measurement—rather than estimations from a representative sample—may become a dominant methodology for both online and over-the-air delivery. Privacy issues play a key role here, so it will be important for such reporting to remain anonymous, and/or require an opt-in by sampled users. At this writing, this remains a controversial and critical subject, which will likely remain a topic of substantial industry concern going forward.

CHAPTER 19

Transmitter Site Facilities

For many reasons, including antenna coverage, zoning, and space requirements, the best location for the transmitting antenna for a radio or television station is often different from the preferred location for the studio facilities, where convenience of access and other considerations may be more important. In that case, the transmitter and antenna will usually be located at a separate site, with the program feed provided by a studio-transmitter link (as described in Chapter 11). On occasions, it may be possible for the transmitter equipment and antenna to share the same site as the studios, in which case the studios and transmitter facilities are said to be *co-sited* or *collocated*.

Viewed overall, transmitter sites for all types of broadcast facilities (AM, FM, and TV, analog and digital) have some similarities. As a rule, most transmitter sites will have at least the following categories of equipment, all as shown in Figure 19.1:

- Studio-to-transmitter link (STL) (see Chapter 11)
- Processing equipment to prepare the signal for transmission
- Transmitter *exciter* to create the RF waveform that will be transmitted (except for AM stations)
- Transmitter *power amplifier*(s) to produce the high-powered signal for broadcast of the exciter's waveform
- *Transmission line* to carry the high-powered RF signal from the transmitter to the emission antenna
- Transmitting *antenna* (with AM, the tower itself is the antenna)
- *Tower* to support the antenna

The exciter and power amplifier together are usually referred to as the *transmitter*.

Although all AM, FM, and TV transmission facilities have a lot in common, the details vary greatly—and even more so when digital IBOC equipment is

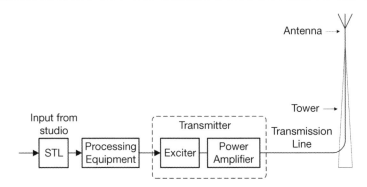

FIG. 19.1.   Transmitter Site Block Diagram

considered. There are also major differences in the types of antennas that are used. First, consider the arrival of the signals at the transmitter site.

INCOMING FEEDS

Unless the studio and transmitter site are collocated, all signals, whether they are audio or video, analog or digital, arrive at the transmitter site via the studio-transmitter link (STL), as detailed in Chapter 11. These feeds may be wireless links, or wired connections (on twisted pair, coaxial cable or optical fiber) provided by third-party telecommunications companies. Both analog and digital STLs are still used, but with DTV the STL has to be digital, and this is also increasingly the case for radio. Where necessary, audio and/or video signals coming out of the STL's terminal equipment at the transmission site may be fed to analog-to-digital or digital-to-analog converters before going to the next links in the chain.

For radio transmission, audio signals out of the STL can either be stereo left and right, mono (for an analog AM station), or composite stereo audio, ready to feed directly to an FM analog exciter. Audio feeds for IBOC digital radio transmission may be taken from the same feed, or separate feeds sent via the same or a different STL, and may be either in the form of stereo left and right audio signals, or a digital signal ready for feeding directly into the IBOC exciter.

In the NTSC era, video signals from the STL were usually composite analog NTSC, but in some cases they were sent from the studio as component digital video that then had to be converted to analog for transmission. (NTSC audio signals were handled like analog FM above—either as discrete left and right audio channels, or a composite signal from the stereo generator.) DTV signals almost always arrive at the transmitter site in the form of a single digital bitstream, ready to feed directly to the exciter.

PROCESSING EQUIPMENT

AM Radio

AM analog stations are now usually mono only; the audio processor needed to optimize the sound of the station may be located either at the studio or at the transmitter site. If the station is also transmitting AM IBOC, and a common STL is used for both analog and digital audio transmissions, the audio processing must be at the transmitter site.

FM Radio

For analog FM radio a *stereo generator* is used to create the composite stereo signal, as described in Chapter 12. The stereo generator has two main inputs—left and right channel audio. It has a single composite baseband output comprised of the mono, stereo, and pilot signals, as shown in Figure 12.3.

For stations carrying ancillary audio services, the SCA subcarrier is combined into the composite audio signal, as shown in Figure 12.4, either in the stereo generator or in a separate SCA device. Radio data system (RDS) information, if provided, is also added here.

Some FM stations place the stereo generator at the studio and feed the composite signal over the STL. Others feed component left and right audio and ancillary services over the STL and place the stereo generator (and other data encoders, if used) at the transmitter site. The second arrangement has to be used if a radio station is transmitting IBOC and analog services using a common STL.

Typically, the output of the stereo generator is fed into the exciter (see below) for modulation onto the station's main carrier. Sometimes the audio processing equipment used to enhance the station's sound includes the stereo generator.

IBOC Radio

For AM-IBOC, separate left and right audio signals are usually sent to the transmitter site for transmission. This can also be done for FM, but a more recent variation allows the FM-IBOC signal to be assembled at the studio and sent via digital STL to the transmitter where it is directly fed to the IBOC encoder. If IBOC-FM audio is sent over a separate STL from the analog service, audio processing to optimize the sound of the station may be carried out at the studio, or before feeding into the IBOC encoder at the transmitter site. The processing for the analog service is often different than that applied to the IBOC service, so if a common STL is used, then the separate processing for IBOC and analog must be applied at the transmitter site. FM radio audio processors today often contain separate processing chains and outputs for analog and IBOC digital signals, so such independent audio processes may be accomplished by a single device.

NTSC Video

In the NTSC era, when the video signal arrived at the transmitter site via an analog STL, it was in a composite video form ready to feed directly to the NTSC video exciter (see below) with little, if any, processing required. If it arrived as component digital video, it had to be converted to analog composite video at the transmitter site before feeding it to the exciter.

ATSC DTV Bitstream

For regular (i.e., fixed) ATSC services, the DTV compressed bitstream sent to the transmitter usually contains the complete and final video, audio and data signals. No further processing needs to be done at the transmitter before feeding to the exciter, except in the case that additional datacasting inputs are added to the bitstream at the transmitter site.

If an ATSC Mobile DTV service (see Chapter 16) is broadcast by a station, however, additional processing is required at the transmission site. This is because the Mobile DTV service is quite different than the fixed ATSC service—using different audio and video codecs, a different transport, and much higher levels of error correction—and it must be added to the broadcast signal in a way that will not disrupt legacy ATSC receivers. Therefore the Mobile DTV signals are processed separately and carefully inserted at a particular point within the ATSC encoding process (which is why this must be done at the transmitter site).

EXCITERS

The *exciter* is the device that takes the baseband audio, video, or digital bitstream baseband signal and converts it to a radio frequency carrier with the appropriate method of modulation. The output of the exciter is at a low power level and has to be amplified further to produce the high power needed to feed the antenna. Sometimes the exciter is installed inside the transmitter housing, but otherwise it is a separate unit installed in an equipment rack near the transmitter.

Analog AM Radio

Most analog AM radio transmitters (except for those transmitting analog stereo) do not have an exciter. Instead, the amplitude modulation process takes place in the transmitter's power amplifier, as described below.

Analog FM Radio Exciters

For FM radio, the exciter takes the incoming composite audio and data signals and frequency modulates them onto an RF carrier at the station's frequency.

AM IBOC Exciters

The AM IBOC exciter takes audio inputs for both the analog and digital services. It modulates the IBOC service onto the digital carriers using QAM and COFDM, and amplitude modulates the analog signal directly onto the station's main carrier. The combined composite output is then fed to a single common power amplifier.

In the exciter, the analog audio is delayed by several seconds to ensure that the analog and digital audio are in sync at the output of the receiver. This is necessary because the digital signal path intrinsically has several seconds of delay (typically about five seconds). Some of the delay is caused by digital signal processing, but most is created by the IBOC *diversity delay* buffer. Diversity delay provides extra robustness in reception of the signal by mitigating problems due to short interference events (e.g., temporary signal loss that can occur with a moving vehicle).

FM IBOC Exciters

Although they may utilize separate exciters for the analog and digital portions of the signal, FM IBOC stations typically use a single combined exciter unit. The combined exciter takes audio inputs for both the analog and digital services. It modulates the IBOC service onto the digital carriers using COFDM. If the station is using a common transmitter/power amplifier for IBOC and FM services (see below), it also frequency modulates the analog signal onto the station's main carrier, after applying the diversity delay similar to AM IBOC. The combined composite output is then fed to a single common power amplifier.

For stations that install a new, separate IBOC transmitter while retaining their existing FM transmitter, the delayed analog audio is fed out of the IBOC exciter and connected to the input of the existing analog FM exciter. In that case, the IBOC exciter composite output feeds only the IBOC power amplifier.

NTSC TV Exciters

In NTSC television, there were two exciters: one for video and one for audio. The video exciter took the incoming composite video signal and amplitude-modulated it onto the station's video carrier. The aural exciter took the incoming audio signal and frequency-modulated it onto the station's aural carrier. These two signals fed either a single or two separate power amplifiers (see below).

ATSC DTV Exciters

For regular (i.e., fixed) ATSC DTV service, a single exciter takes the ATSC data stream and modulates it onto the station's carrier, using 8-VSB modulation.

If ATSC Mobile DTV service is broadcast by the station, however, a different type of ATSC exciter is required, which processes the fixed and mobile services separately, then combines them for transmission in such a way that legacy ATSC receivers will ignore the Mobile DTV data.

POWER AMPLIFIERS

The second main part of a broadcast transmitter is the *power amplifier*. This takes the modulated RF waveform from the exciter and amplifies it to the high-power radio frequency signal needed to drive the broadcast transmission antenna (for traditional analog AM transmitters, the arrangements are somewhat different—see below). Power amplifiers for analog and digital transmission generally use similar components, although the way they are set up and used may be very different. It may be possible to convert some (but not all) recent models of transmitters from analog to digital operations. *Transmitter power output* (TPO) may vary from a few hundred watts (W, the unit of electrical power) to tens or even hundreds of kilowatts. An example of a high-power UHF television transmitter is shown in Figure 19.2.

FIG. 19.2.   UHF Digital Television Transmitter. Courtesy Lawson & Associates, DesignTech, and KDTV

Tubes and Solid-State Devices

The heart of many power amplifiers is one or more high-power electron tubes. There are various different types of tubes, each of which work best for particular wavelengths and output powers. The most common transmitting tube devices are *tetrodes*, *klystrons*, and *inductive output tubes* (IOTs), but there are others.

For various reasons, all electron tubes wear out. They typically have to be replaced every few years, although this varies greatly for different transmitter and tube types. To get away from the tube-replacement routine, many modern transmitters have been designed using all-solid-state electronics. Solid-state power amplifiers use the same principles as transistors and integrated circuits, and are very reliable. They do not require periodic replacement like tubes, although some high-power solid-state amplification components can still fail and require replacement on infrequent occasion.

Despite their high reliability, solid-state electronics are not generally used for the highest powered transmitters, such as those used in many UHF TV stations, because they are less efficient than tubes and, at high powers, would result in higher electrical power costs.

Cooling Systems

The tube or solid-state components in a power amplifier get very hot during operation and must be kept cool to prevent overheating damage. This is performed either by blowing cold air through the transmitter, or with a liquid cooling system akin to that used for cooling the engine on a car, using a radiator outside the building to dissipate the heat. Providing adequate cooling occupies a considerable amount of the design and operation of a broadcast transmission site.

AM Radio Amplification

Traditional analog AM broadcast transmitters are fundamentally different from most other transmitters. The RF input to their power amplifiers is typically an unmodulated carrier wave from a device called an *oscillator*. After first being amplified, the incoming program audio signal is also fed to the power amplifier; the amplitude modulation process for the carrier then takes place in the power amplifier.

FM Radio Amplification

Analog FM amplification is fairly straightforward, with the FM exciter output fed to one or more stages of power amplification. Because the FM waveform generally remains at the same amplitude (i.e., only its <u>frequency</u> varies with modulation), relatively simple and cost-effective amplifiers can be used. The power levels and other parameters of FM radio transmission in the U.S. lend themselves well to solid-state transmitter designs.

AM IBOC Amplification

Because AM IBOC requires the analog and digital signals to be modulated together (a process known as *low-level combining*), they have to share a common power amplifier. It may be possible to upgrade some more recent analog AM transmitters for this purpose, but frequently a new transmitter is required when converting from analog only to hybrid AM IBOC operation.

FM IBOC Amplification

FM IBOC has four alternative transmission arrangements: low-level combining, *mid-level combining*, *high-level combining*, and *dual-antenna*. In low-level combining, the analog and digital signals are combined together in the exciter and then share a common power amplifier (as in AM IBOC). In high-level combining, the analog and digital signals have separate power amplifiers, and the outputs are combined in a *diplexer* (see later section). Mid-level combining is a combination of the low- and high-power combining techniques and may be advantageous for stations with moderate (on the order of 10kW) power outputs. The dual-antenna system has two completely separate power amplifiers and antenna systems, one for the analog portion of the hybrid IBOC signal and one for the digital, which must be located close to each other (i.e., on the same tower, or on two adjacent towers at the same transmission site).

NTSC TV Amplification

During the NTSC era, VHF analog television stations were typically configured with the video and audio exciters feeding their output signals into two separate power amplifiers. Then the outputs from the two power amplifiers were combined and fed via a single transmission line to the antenna. On the other hand, many UHF station transmitters mixed the audio and video signals together within the transmitter, sharing a common final power amplifier stage. The transmitted audio power was between 5 and 10 percent of the video power.

ATSC DTV Amplification

DTV uses a single power amplifier (although it may be composed of multiple stages) to boost the power of the digital bitstream's waveform created in the ATSC exciter. In some cases, it may be possible to convert more recent models of analog TV transmitters to carry a digital signal.

TRANSMISSION LINES AND OTHER EQUIPMENT

The *transmission line* (also known as a *feeder*) that connects the transmitter or diplexer (see next section below) to the antenna is generally either a length of *flexible* or *semi-flexible coaxial cable*, *rigid coaxial feeder*, or *waveguide*, with the choice dependent on the frequencies and power levels involved. Another type called *open wire* feeder is occasionally used, mostly for shortwave stations.

Flexible or semi-flexible coaxial transmission line is like an extremely thick piece of cable television coaxial wire. Its thickness depends on the amount of power it has to handle from the transmitter (if it is too thin it can overheat, or spark over during high voltage peaks), ranging from about an inch to several inches in diameter. Nonflexible types of transmission lines are made of rigid copper pipe sections. They have a narrow section of pipe threaded through the middle of a wider section of pipe, with the narrow inner pipe kept exactly centered within the larger outer pipe by installing plastic spacers inside the larger pipe. Rigid coaxial lines, as their name implies, cannot be bent, and like water plumbing, special angled sections are needed to change the direction of the line. They can be used for very high power antenna feeds, with sizes up to about 12 inches in diameter, the exact size required again depending on the signal power and frequency.

Waveguides are best described as "duct work for radio frequencies" and do not have a center conductor or any of the insulating material associated with it. In fact, they look much like the metal ducts you might find in an air-conditioning system. Radio waves propagate down waveguides by reflecting off the inside of the metal walls of the guide. Waveguides are the most efficient type of broadcast transmission line; that is, more of the transmitter's energy makes it to the antenna without being lost as heat compared to other types of transmission lines. The main disadvantage of waveguide is that it is considerably larger than coaxial transmission line, and therefore it is more difficult for a broadcast tower structure to support. It is most often used with high-power UHF transmitters, where the short RF wavelengths keep the waveguide size manageable, and its superior efficiency helps save a significant amount of electric power.

Dummy Loads, Diplexers, and Reject Loads

It is rare for a broadcast transmitter to be connected directly to the antenna by the transmission line. Usually, there is some sort of large switch or connection panel between the transmitter and antenna that allows the transmitter signal to be fed to one of several destinations, including a *dummy load* in the transmitter building. The dummy load accepts the transmitter's RF power and turns it almost completely into heat, allowing the transmitter to be tested without radiating over the air. Similar RF switching panels may be used to direct a transmitter's output to one of several antennas that may be available to it (see section on Backup Systems below).

There are also cases where the outputs of two or more transmitters need to be fed to a single antenna, for reasons that include the following:

- Combining multiple radio or TV transmitters operating on different channels into a shared antenna
- High-level combining of FM IBOC and analog transmitters' signals
- Combining multiple transmitters for redundancy purposes
- In the NTSC era, high-level combining of audio and video carriers
- During the DTV transition, combining a station's NTSC and DTV transmissions

In these cases, the transmission lines from two transmitters are connected to a *diplexer* that allows their signals to be combined together. In cases where more than two transmitters are involved, the device is called a *multiplexer*. Both of these device classes are often collectively referred to as *RF combiners*. In order to balance the RF combining system, it is often necessary to feed a portion of the signal to a *reject load* rather than the antenna; this load turns the unwanted signal into heat, as with a dummy load.

Combining RF signals is complex and not always possible or practical. Whether or not multiple transmitters can be combined to feed one antenna depends on many factors, largely related to the design of the antenna, but also dependent on the transmitter frequencies and output powers.

The FM IBOC high-level combining application is a particularly challenging case, in which both transmitters occupy the same frequency, and one is at substantially higher power than the other.

AM ANTENNA SYSTEMS

Broadcast antennas differ greatly from one type of broadcast service to the next—AM, FM, or TV. The main reason is that, in order to operate efficiently, the length of any antenna (whether used for transmission or reception) must be related to the *wavelength* of the transmitted signal. For efficient operations, generally the antenna will be somewhere between 1/4 wavelength and 1/2 wavelength long. As noted in Chapter 7, radio waves used for broadcasting have a wide range of wavelengths, ranging from one or two feet for UHF television waves to hundreds of feet for AM radio waves, and thus antenna size also varies greatly.

AM Antennas

The wavelength for AM transmissions is several hundred feet, with the exact wavelength dependent on the station's carrier frequency. Therefore, an AM radio station's transmitting antenna is usually the station's tower itself. For this reason, AM towers are considered to be *hot* (i.e., energized with electrical energy from the transmitter), and they can severely shock or kill a person who touches the metal structure of the tower while standing on the ground. To avoid such injuries, the FCC requires this type of tower to have a fence and proper signage around it.

The height of an AM tower depends on the transmitting frequency of the station and whether it is a 1/4-wave or 1/2-wave antenna. Generally, the lower the station's frequency, the taller the tower, with a maximum of about 450 feet.

A single AM tower will have a non-directional horizontal radiation pattern, transmitting the same amount of energy in all directions, as shown in Figure 19.3.

Directional Arrays

In some cases, multiple AM radio towers are used together as part of a *directional antenna* (DA) system, also called an *array*. An AM directional antenna array involves two or more transmitting towers standing in a straight line, or sometimes arranged in a parallelogram formation. The purpose of the DA is to

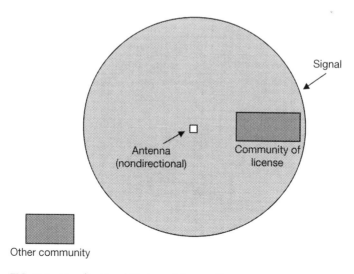

FIG. 19.3. Non-directional Horizontal Antenna Pattern

direct the transmitted energy toward the community of license and to reduce the energy traveling in directions where it might cause interference to other licensed stations. This is done by varying the phase of the transmitted signal among the several towers. Generally speaking, the larger the number of elements in the array, the more highly directional the pattern. Figure 19.4 shows an example of a directional *horizontal radiation pattern*. A directional antenna is said to have a certain amount of *gain* in the direction of most power, when compared to a non-directional antenna, and this increases the *effective radiated power* (ERP) of the station, which is frequently much larger than the TPO. ERP is simply calculated by multiplying TPO (in watts) by the gain of the antenna (in dB). For example, a transmitter operating at 10 kW TPO fed to an antenna with a gain of 5 dB results in an ERP of 50 kW.

In many instances, AM stations are authorized by the FCC to operate in a non-directional mode during the day but are required to use a directional mode at night. The reason for this is that an AM transmitter that does not interfere with signals in other communities during the day may do so at night as a result of *skywave* propagation. For more information about skywave propagation and the FCC's methods for preventing interference with other broadcast stations, see Chapter 20.

Ground Radials

Another important component is the antenna *ground radial* system. AM radio waves traveling across the surface of the earth (the *ground wave*, see Chapter 20) depend on *ground conductivity*. They need an excellent connection to the ground at the transmitter site to give the signal a good start as it leaves the transmitter. This is achieved with a series of ground radials, which are copper wires buried in the ground, extending outward from the base of the antenna for several hundred feet in a pattern like that shown in Figure 19.5. A standard AM ground

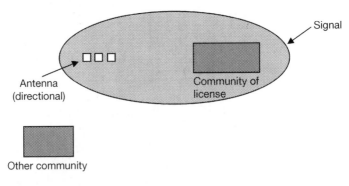

FIG. 19.4. Example of Directional Antenna Pattern

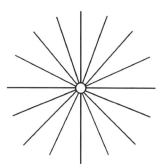

FIG. 19.5.   Overhead View of AM
Ground Radials (Circle at Center
Represents Base of Antenna
Tower)

radial system will have many more equally spaced radials than shown in the drawing (typically 120), and the actual length of each ground radial is a function of the transmitting frequency of the station–the lower the transmitting frequency, the longer the radial.

Because of their dependence on ground conductivity, AM radio transmitting antennas are usually found near the communities they are intended to cover, at approximately the same site elevation.

AM IBOC Antennas

Many antenna systems used for analog AM broadcasting should also be able to carry AM IBOC signals. In some cases, it may be necessary to adjust some parameters on the antenna and/or the transmission feed line to optimize performance of the system. Most important is to provide as wide a bandwidth as possible through the antenna system, which is important because IBOC carriers are largely concentrated at the outer edges of the channel (as shown in Figure 13.1). This is particularly challenging on AM directional arrays.

FM AND TV ANTENNAS

Wavelengths in the FM and TV broadcast bands range from just over a foot for the highest UHF TV frequencies to about 10 feet for FM radio and low-band VHF television. Although antenna designs do vary somewhat for FM and TV, the concepts are not very different, so they are covered together below.

FM and TV transmitting antennas are usually found near the communities they are intended to cover and are mounted as high as possible, usually on a tall tower. Where it is practical, the tower is located on the highest terrain in the area, often a mountain where available. This is because the short wavelengths of FM and TV band signals tend to travel in straight lines and do not "flow" around obstacles. They therefore perform best when there is an unobstructed path between the transmitter and receiver (i.e., line-of-sight). In addition, they are far less dependent on ground conductivity for propagation than AM band waves are.

Also because of their short wavelength, FM and TV antennas are much smaller than AM antennas, typically ranging from somewhat less than a foot to about five feet long. They are electrically isolated from the tower on which they are mounted, which is grounded and consequently safe to touch. The antennas are often made up of multiple *elements*, which affect the directional pattern and ERP.

The antenna elements are mounted on their supporting structure with clamps or other mounting hardware, and the complete antenna assembly is then attached to the supporting tower. The antenna is fed with the signal from the transmitter by the transmission line that extends from the transmitter up the tower to the antenna.

Multiple Bay Antennas

FM or TV antennas are often referred to as "single bay," "2-bay," "4-bay," and so on. The word *bay*, in this case, means a single antenna element. Multi-bay antennas are often used in FM radio and TV transmission because they make the transmission system more efficient by changing the *vertical radiation pattern* to focus the beam and reduce the amount of energy sent up into the sky, and sometimes to reduce the amount of signal that is transmitted toward the area in the immediate vicinity of the tower, which would otherwise receive more signal than it needs. The more bays that are added to an antenna system, the more focused the transmitted signal becomes in the vertical plane. For example, Figure 19.6 shows how the transmitted signal from a 2-bay antenna might differ from the transmitted pattern of an 8-bay antenna.

A multi-bay FM or TV antenna operates using the same principles as an AM directional antenna (DA). The major difference (aside from the frequency, and therefore the size of the antenna elements) is that the multiple antennas in an FM or TV array are stacked vertically on the tower, whereas the multiple antennas in an AM DA are lined up horizontally along the ground. The result is that the FM or TV multi-bay antenna focuses its transmitted energy in the vertical direction, whereas the AM DA focuses its energy in the horizontal direction. As with AM DAs, multi-bay FM antennas have higher gain than single-bay antennas, and this gain is proportional to the number of bays in the antenna. Again, higher ERP results from multi-bay FM antennas, calculated as described in the section on AM Directional Arrays above (TPO × Antenna Gain = ERP).

It should be noted that in some instances it is necessary to focus an FM or TV signal in a particular <u>horizontal</u> direction (like AM DAs), as shown in Figure 19.4.

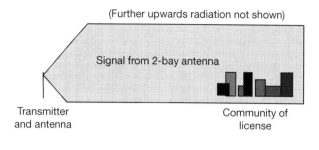

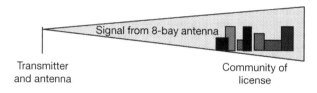

FIG. 19.6. Effect of Multiple Antenna Bays on Vertical Radiation Pattern

Therefore some FM or TV antenna systems are designed for these purposes, using concepts similar to those described for AM previously—although the smaller wavelengths used for FM and TV allow such patterns to be formed using a single directional antenna array on a single tower. Pattern adjustments can be made electronically by adjusting the *phase* (timing relationship) between the signals that arrive at each of the antennas in an antenna array. In this way, it is possible to create complex patterns to cover particular target areas while protecting other directions from unwanted signals. Side mounting an antenna on a tower can also cause some horizontal directionality, which can be used to advantage when a directional pattern is required, or compensated for in the antenna design and mounting when an omnidirectional pattern is desired from a side-mounted antenna.

FM IBOC Antennas

Most antenna systems used for analog FM broadcasting should be able to also carry the FM IBOC signals without significant adjustments. In the alternative dual-antenna arrangement for FM IBOC, the IBOC signal from the digital transmitter is fed to a separate second antenna, which must be in close proximity to the first, as mentioned above. In that case, the analog transmitter and antenna system are left unchanged.

Combined Channel Antennas

It is possible to combine the outputs of two or more transmitters on different channels together and feed the combined signal up the tower to a single *wideband* antenna. This is most often done where tower space is at a premium (e.g., in large cities such as New York). It is also done elsewhere where there are good reasons to do so, and can result in considerable savings in space and overall cost.

Most antennas operate most efficiently over a narrow range of frequencies, but special designs are available to handle multiple channels. The channels to be combined have to be within a certain range, because the wider the range of frequencies to be handled, the greater the compromises in antenna performance that have to be accepted.

TOWERS

Towers are of two main types: *self-supporting* and *guyed*. The latter (also known as *masts*) are held up with guy wires, which are usually metal cables—although sometimes they are made from a special type of nonmetallic rope—attached to anchors in the ground. Self-supporting towers are usually in the range of about 1000 feet or less, and guyed towers can reach up to about 2000 feet.

It is also common for broadcast antennas to be mounted on short structures atop existing tall buildings, particularly in city locations with skyscrapers. This avoids the cost of a separate tall tower. Special provision has to be made to house the transmitter equipment in a suitable space within the building, and care must be taken that occupants of the upper floors of the building or adjacent buildings are not subjected to excessive RF exposure.

As previously mentioned, AM towers usually form the actual transmitting antenna. For all other broadcast services, however, the tower is used to <u>support</u> the antenna, which is either fastened to the top of the tower (*top-mounted*) or to the side of the tower (*side-mounted*). Antennas may be open to the elements or protected inside a fiberglass *radome*. One tower frequently supports antennas for more than one transmitter, including FM radio and DTV services, so multiple stations may often share the same tower, along with other, non-broadcast services such as cellular telephony, two-way radio for industrial and emergency-service providers, and other users. In some cases, antennas for FM radio, cellular telephony, and other services can be mounted on an AM antenna (i.e., tower), which gives an AM antenna owner the opportunity to make additional revenue by leasing tower space. Such mounting must be done carefully, however, as to not disturb the propagation and coverage pattern of the AM station.

Markings and Tower Lights

According to FCC and Federal Aviation Administration (FAA) regulations, towers over 200 feet high have to be painted in red and white stripes and/or fitted with lights to help indicate the location of the tower to aviation traffic.

Wind Loading and Icing

The tower has to be designed to safely withstand the strongest winds that occur at the site, taking account of all the extra loads imposed by the antennas and transmission lines. This is important to remember when changes or additions to antenna systems are planned. Quite frequently, tower strengthening may be required when new antennas are added to an existing tower.

Depending on the tower location, accumulations of ice on the tower and antennas in cold weather may be a major issue, affecting the design and construction of the tower. De-icing heaters are commonly installed in extreme climates. Tower failures (i.e., partial or complete collapses) rarely occur, but when they do, icing is a common cause.

TRANSLATORS AND REPEATERS

As mentioned earlier, FM and TV radio signals tend to travel in a line-of-sight fashion and do not flow well around obstacles. This means that in regions with uneven terrain, there may be population centers that, while within the intended service area of the station, are in a valley or behind a ridge and completely hidden from the view of the transmission antenna, and therefore unable to receive the service. To overcome this problem, a *translator* can be added, spectrum permitting. Translators use a receive-antenna and high-quality FM or TV receiver at a suitable high location that can receive the main station. The output of the receiver is converted (*translated*) directly to a different RF frequency (often without converting it to baseband audio or video) and retransmitted, often with a directional antenna, toward the area where it is needed. These translators typically operate at low powers between about 1 and 250 watts.

In some cases, the fill-in transmitter operates on the <u>same</u> frequency as the main transmitter and is then known as a *repeater* or a *booster*. Again these are typically low-power devices, but some stations may be licensed to operate high-power repeaters when terrain and population distribution warrant. In other cases, stations may use multiple translators and/or repeaters in a "daisy chain" arrangement to carry the signal onward to remote locations.

TRANSMITTER REMOTE CONTROL

The majority of broadcast transmitters in operation today are connected to a remote control unit. For many years, the most popular remote control systems have used two hardware devices, one located next to the transmitter and one at the remote control point, which is usually the master control room at the studio site of the station operating the transmitter. The two ends of the remote control system are connected by a communications link, which may be a dial-up or dedicated phone line, a communications channel on the studio-transmitter link (STL), or some other telecommunications path. The units communicate with each other via data communications similar to computers communicating via the Internet. In some cases, one or both of the remote control devices may in fact be PCs running transmitter remote control software.

The unit at the transmitter site has a series of *control contacts* that are connected by cables to the switches that control the transmitter and optionally, the operation of other devices at the site. The transmit-site unit also has connections that monitor the status of the transmitter (e.g., on or off, alarms and fault conditions, various voltage and power readings), as well as other optional status indicators (e.g., indoor and outdoor air temperatures). A complementary unit at the studio location has switches and indicators that automatically mimic those at the transmitter; they allow an operator to control and monitor the transmitter (and other devices or conditions) as though the operator were physically at the transmitter site.

The number of controls and indicators will vary greatly from system to system. In addition to interfacing to the transmitter and other related equipment, most remote control systems also include other connections for site functions, such as electric power conditions, security alarm status, and in particular, monitoring and switching of a directional antenna, other RF routing control, and tower-lighting status. Remote control systems are usually able to automatically print out the values of the various transmitter meter readings that are required to be recorded by the FCC. These become part of the station log.

Remote Dial-up

Some systems allow a transmitter engineer to call the remote control unit from any touchtone phone and control the switches in the remote control unit—and, by extension, the switches on the transmitter—by pressing buttons on the telephone. The transmitter may also report its status with synthesized voice messages over the phone.

Computer-Based Remote Control

Modern remote control systems have computer monitors and keyboards at the studio control point, which allow for a better human interface and increased capabilities. More sophisticated still are *network monitoring* systems, which often work over computer networks (including the Internet) and are able to provide complete management of transmitter sites and other facilities, potentially from any PC (password protection is used to prevent unauthorized access or tampering).

Modern transmitters usually offer remote control connections like the serial port or Ethernet connection on a computer. These do not require dedicated hardware switching and can connect directly via the communications link to the studio control point.

BACKUP SYSTEMS

All of the above systems can be supplemented by backup systems that provide redundancy for each component of the signal path, so no single point of failure will cause a station to curtail its services. For example, it is not uncommon for a station to have a backup STL, backup processing equipment and encoders, backup transmission lines, transmitters and antennas, and even complete backup transmitter sites, along with backup power generators at each site to prevent nearly any instance of a failure taking the station off the air. An additional element of any backup system is the ability to quickly (and in many cases, automatically) switch from any primary system to its backup, which adds considerable—but worthwhile—complexity to a station's signal path and control systems.

CHAPTER 20

Radio Wave Propagation and Broadcast Regulation

This chapter discusses aspects of the Federal Communications Commission's (FCC) rules that regulate broadcasting in the United States, and considers some international issues having impact on broadcast regulation. It begins with a discussion of the propagation characteristics for the different broadcast frequency bands, which influence the rules for different types of stations.

Propagation characteristics are the particular qualities of a signal that determine how it behaves as it travels through the atmosphere (i.e., how far and in what direction). As mentioned in Chapter 7, because of their different wavelengths, the carrier waves used for AM, FM, and TV transmissions have different properties for how they propagate through the air. What follows is an exploration of the main characteristics of broadcast RF propagation, with some specific considerations for digital transmissions.

AM PROPAGATION

Skywave

What makes an AM signal so different from an FM or TV signal is the need to consider *skywave* propagation, which affects the MF band used for AM radio—but does not affect the VHF and UHF bands used for FM radio and television. *Skywave* is a phenomenon that occurs after sunset, when the absence of sunlight causes changes to occur in the Earth's upper atmosphere, making it reflective to longer radio wavelengths. Therefore signals that would pass through or be absorbed by the atmosphere during the day instead bounce off of the *ionosphere* layer at night, and return to the surface of the Earth at distances far from the transmitter, as shown in Figure 20.1.

As a result, an AM skywave signal is capable of traveling from its transmitter to a distant city, perhaps skipping over many places in between. For this reason, care

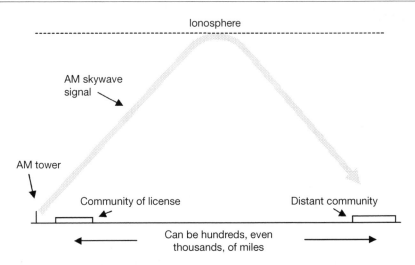

FIG. 20.1. AM Skywave Propagation

must be taken when allocating an AM frequency to a new station to ensure that the transmissions at night will not cause unacceptable levels of interference because of skywave propagation to distant AM stations on the same or adjacent channels.

Because of the need to protect other AM stations from skywave propagation, the FCC allocates some AM frequencies to stations on a daytime-only basis. These stations are generally called *AM daytimers*, and they must turn off their transmitters at night. In other cases, AM broadcasters are licensed to operate at one power and/or antenna pattern during the day, and at a lower power and/or more restricted directional pattern at night. Because daytime hours are seasonally variable in most U.S. locations, the time of day at which these changes in operational parameters must be implemented by affected AM stations shift, usually on a monthly basis.

Groundwave

FCC rules are also intended to protect broadcast stations from each other's *groundwave* signals, which remain constant for MW frequencies during both daytime and nighttime.

The groundwave signal, as the name implies, travels over the earth's surface at ground level. How far the groundwave signal travels, for a given transmitter power and antenna, is largely a function of the Earth's *ground conductivity* over the route being traveled by the signal. The higher the conductivity, the better the AM signal travels. Ground conductivity is a measure of the resistance of the ground to electrical signals, which varies considerably in different geographical regions, as shown in Figure 20.2 (based on a map in section 73.190 of the FCC rules).

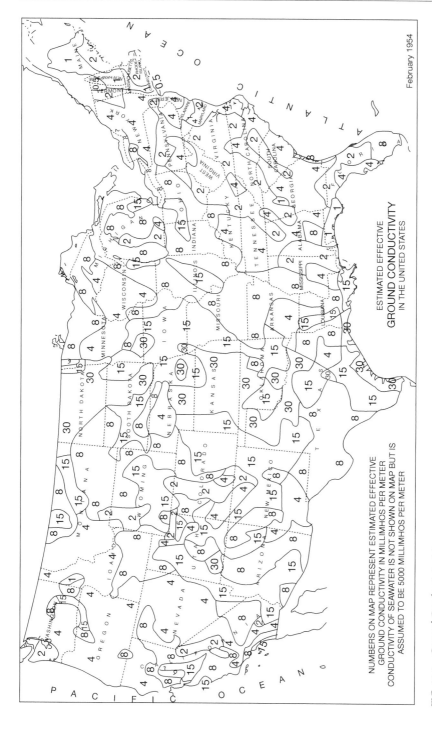

FIG. 20.2.   Ground Conductivity Map

FM PROPAGATION

There are basic differences between the way a MW groundwave and a VHF (the band used by FM radio) signal propagate. The FCC's rules for allocation of FM radio channels are, therefore, very different from the rules for AM channels. For FM radio, neither skywave nor groundwave propagations are issues. The main considerations are line-of-sight and distance.

Line-of-Sight Propagation

Generally speaking, as radio frequencies get higher, they become increasingly dependent on having a line-of-sight view between the transmitter and the receiver. FM-band signals tend to have this characteristic, and their propagation is generally not dependent on the conductivity of the ground they are traveling over.

To understand this point, think about visible light, which has frequencies much higher (and therefore wavelengths that are much shorter) than broadcast radio waves. Light has a hard time getting around any obstruction, and casts a hard shadow. AM-band waves, however, with wavelengths in the hundreds of feet, are able to follow the surface of the earth and largely "flow" around obstacles or even mountains. FM-band waves, with shorter wavelengths than AM, come somewhere in between, but they perform best in an environment with no obstructions.

The FCC's frequency allocation procedure for FM radio stations takes into account factors including the general layout of the land around the proposed antenna site and the distance between the proposed site and other stations on the same and adjacent channels. For FM stations at the lower end of the FM band, the distance to the nearest TV channel 6 transmitter is also taken into account, because channel 6 is just below the lower edge of the FM band.

Multipath

Another characteristic of the shorter wavelength FM signals is their tendency to reflect off the earth's surface when they encounter an obstruction. They will also reflect off objects such as buildings and large vehicles (much as light reflects of shiny surfaces). Therefore, an FM transmitter located in an area where there are mountains or large hills may have major propagation problems for two reasons. First, the signal from an FM transmitter located on one side of a mountain will have a hard time reaching the other side because the mountain will block the primary signal. Also, there may be reflections off one or more hillsides, producing some transmitted energy in a different direction from that in which it was originally traveling.

This results in a phenomenon called *multipath interference,* which, as the name implies, occurs when FM receivers get signals from the same FM transmitter via more than one path. An illustration of this concept is shown in Figure 20.3.

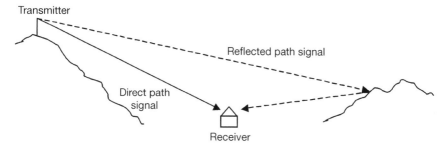

FIG. 20.3.   Illustration of Multipath Interference

Although the two received signals are identical in content, the longer path traveled by the reflected signal will cause the two signals to be slightly *out of phase* (i.e., the reflected signal arrives at the receiver a small fraction of a second after the direct signal). Because of this, the reflected signal is often called an *echo*.

The phase difference of the two signals causes them to interfere with one another. In another example, multipath manifests itself as the "picket fencing" effect that motorists may notice on their FM car radios as they move slowly in a city street. The picket fencing, or repetitive noise and distortion, is caused by two or more multipath signals bouncing off buildings near the street. As the car moves, the reflected signals alternately add and subtract from the main signal, causing *frequency-selective fading*, which results in the audible interference.

IBOC CONSIDERATIONS

The system design for IBOC is intended to approximately replicate the primary service area of the analog channel with the digital program service, as described in Chapter 13. Although in its hybrid mode the IBOC digital signal coexists in the same channel as a station's analog transmission, the digital signals are gener-ated as a wholly separate service—in some cases using a separate transmitter, or even separate antennas. For these reasons, propagation and reception capa-bilities may differ in actual practice between the two services of a given station. Moreover, it is well known that the gradual degradation of analog transmis-sions allows some users to tolerate listening at greater distances and lower signal strengths, accepting the increased noise and distortion that may occur. IBOC digital service may not be receivable under these conditions. It was also consid-ered necessary for the primary digital service ("HD-1") to degrade gracefully at the edge of the service area, so the listener would not suffer the "cliff effect." Therefore the system was provided with the capability to blend more-or-less seamlessly from the primary digital to the analog service at the receiver, when the digital signal strength became too weak to properly receive.

The IBOC system, using OFDM techniques, was specifically designed to resist multipath interference, which therefore does not cause the audible degradation described above, and normally does not cause IBOC reception to fail.

TV VHF AND UHF PROPAGATION

TV signals in the VHF and UHF bands propagate in basically the same manner as FM signals, but UHF signals are somewhat more directional and dependent on line-of-sight transmission owing to their higher frequency.

Multipath

Multipath interference, discussed earlier with regard to FM signals, can also cause difficulties with TV reception. During the NTSC era, the received audio signal could suffer a similar fate to FM radio under multipath conditions. More commonly, multipath also caused the *ghosting* effect in received analog TV video. The multiple iterations of an NTSC video signal arriving at the receiver at slightly different times due to multipath caused additional images (ghosts) to be displayed on the screen, which were displaced laterally due to their slightly delayed arrival.

On the other hand, ATSC (digital) pictures never display ghosts because of the way the picture is carried as digital data. As mentioned in Chapter 15, the DTV receiver is able to cancel out the unwanted echoes using circuitry called an *equalizer*, and the DTV signal contains special forward error correction codes that allow the receiver to correct for errors in the received signal. Under extreme conditions, however, low signal strength and/or large multipath echoes can cause the DTV receiver to be unable to correctly decode the picture (and/or the sound), and reception will fail.

Antenna Height

An FM or TV signal is able to "see" further, and therefore propagate over longer distances, if its transmission antenna is raised up higher. Therefore, if an antenna is raised to a new location from its originally authorized height, the FCC rules require the broadcaster to lower the station's authorized transmitter power. Conversely, if the antenna is moved to a lower location, then the transmitter power may be increased. Such an antenna rearrangement may occur, for example, when a new antenna is added to a tower, and the rules help ensure that moving the antenna does not cause new interference to other authorized stations.

Because digital signals are inherently more robust than analog transmissions, they are usually transmitted at a much lower power level than an equivalent analog transmission on the same channel and antenna height. Although, for various reasons, replication may not occur exactly throughout the service area, when transitioning from analog to digital service, the aim is usually for the digital service to replicate as closely as possible the service area of the existing analog transmitter, and this can typically be done with lower transmit power on the digital service.

SPECTRUM ALLOCATION

The electromagnetic spectrum is a scarce and critical resource, and its use must be carefully managed. It has become customary for such management to be handled by governmental institutions in most of the world.

The Federal Communications Commission

In the U.S., the Federal Communications Commission (FCC) is charged with the process of electromagnetic spectrum management—along with certain other elements of telecommunications governance—as specified by the U.S. Congress in the Communications Act of 1934, and many subsequent statutes. The FCC is an independent agency of the U.S. federal government, meaning that unlike some other regulatory agencies, it is not part of any Executive Branch department or Cabinet Secretary's domain, but rather reports directly to the U.S. President, who appoints its Commissioners.

Insofar as broadcast operations engage the use of electromagnetic spectrum, they fall under the jurisdiction of the FCC—as do most other users of the spectrum in the U.S. As such, broadcasters are just one of many industries affected by FCC regulations.

Besides administering current usage of the electromagnetic spectrum, another process that can affect broadcasters is the FCC's role in managing transitions, which it is charged to do in a strategic fashion that maximizes the social, technological, economic and public-safety benefits of spectrum use in the U.S. This typically involves reallocation of the use of certain parts of the spectrum to other applications, to accommodate the changing needs of users, advances in technology, development of new services, growth or expansion, and other dynamics of modern society. Like urban renewal, this is a complex and multifaceted process with potentially wide-ranging impacts on many stakeholders, and one that can require long periods of time to complete. The U.S. transition from analog to digital television—which began more than a decade prior to its completion in 2009—is a good example of this kind of process, which was managed by the FCC but involved the coordinated efforts of broadcasters, professional and consumer electronics manufacturers, telecommunications providers, retailers, and other government agencies.

FCC RULES

The FCC's technical rules have the primary objective of ensuring that users of the electromagnetic spectrum do not cause unacceptable amounts of interference to other spectrum users. Broadcasters are among these users, of course, and are thus subject to relevant FCC rules. A primary tool used by the FCC is the issuance of *licenses* to authorized parties for their use of a portion of the spectrum under specified parameters. These parameters typically include an operating frequency and channel bandwidth, power level, antenna location, emission type, and other constraints in some cases. As with other licenses, they may be issued for certain periods of time and require periodic renewal, or may be revoked. Another common governance tool used by the FCC to enforce compliance with its rules and licensed parameters is the issuance of monetary fines for violations.

With regard to broadcast operations specifically, another objective of the FCC rules is to ensure that the signals transmitted by the nation's broadcasters use specified modulation types, power levels, and certain other technical parameters that if not properly maintained would be capable of causing interference to other broadcasters, or to other users of the electromagnetic spectrum. The FCC also at one time also regulated the technical quality of broadcast content, but it eliminated this practice in 1984, as part of in a major deregulatory initiative, deciding to leave signal-quality issues to the "forces of the marketplace." A recent and notable exception to this trend was the CALM Act (discussed in Chapters 9 and 16), but that was done pursuant to instructions from Congress rather than on the FCC's own initiative. Certain other elements of the broadcast industry's processes remain under FCC jurisdiction, however, as explained below.

On occasion, the FCC will create new rules that may affect incumbent spectrum users and others (such as when engaging in a transition process as described above). In this case, another common regulatory process called *Rulemaking* is invoked, by which the FCC may poll the industry and public at large on its opinions about possible new regulations action by issuing a *Notice of Inquiry*. As it moves toward a decision on new rules, it may issue one or more *Notice(s) of Proposed Rulemaking*, to which public and industry comment is invited, followed by the eventual declaration of new rules that are about to go into effect via a *Report and Order*. Lesser matters may be handled with the simpler *Public Notice*, stating that certain new regulatory actions have been or will soon be taken.

Broadcast Frequency and Power Allocations

In order to ensure that stations do not interfere with one another, the FCC has adopted complex rules regarding the allocation of broadcast frequencies and power to each station. These rules are based on the propagation characteristics of the broadcast signals involved, as detailed above.

The FCC rules and frequency allocation procedures require different broadcast signals on the same or adjacent channels to be geographically separated by appropriate distances. Different rules also require that each broadcaster's signal must stay in its allocated part of the radio spectrum. These are specified in the FCC's emission masks and modulation limits, which help ensure that one station's signal does not bleed over into the channels of other stations located next to it on the dial.

As noted earlier, AM radio stations have distinctly different propagation characteristics from FM and TV stations, and the FCC has therefore adopted somewhat different allocation procedures for AM stations than for FM and TV. The propagation characteristics of TV and FM signals are similar, so the FCC uses a substantially similar procedure for determining whether a proposed TV or FM channel can be allocated to a particular station.

These differences of propagation are important because they affect the coverage that is achieved from a particular site, the transmitter power that is needed, and the potential interference into other stations on the same or adjacent channels.

Broadcast Auxiliary Spectrum

Beyond the oversight of their broadcast channels, broadcasters also fall under FCC regulation for their other, non-broadcast uses of spectrum (i.e., the wireless, remote-backhaul and facility-interconnection links discussed in Chapter 11). In FCC parlance, this is referred to in general as *Broadcast Auxiliary Service* (BAS), with specific allocations for Studio-to-Transmitter Link (STL), Transmitter-to-Studio Link (TSLs), Remote Pick-Up (RPU), and Electronic News Gathering (ENG) applications. Licenses to use spectrum for STL and TSL purposes are uniquely granted to individual stations, while RPU and ENG spectrum is often shared by broadcasters in a given area, with usage coordinated among the users themselves, or designated to a frequency coordinator in the area—usually one or several station engineers from the area, often operating under the auspices of a local chapter of the Society of Broadcast Engineers (SBE).

Because of the limited number of users of this spectrum—compared to consumer-facing services, for example—BAS spectrum has been subject to more frequent reassignment, under the assumption that transition to new equipment and channels will not have wide-ranging impact. Nevertheless, for broadcasters these changes can be expensive and labor-intensive. They often result in less spectrum remaining available for such usage, as well, so broadcasters have had to be vigilant in their engagements with the FCC to sustain adequate BAS spectrum resources.

Licensed vs. Unlicensed Operations

All of the above uses of spectrum have assumed the need for an FCC license to enable authorized use of electromagnetic spectrum. There is another class of spectrum usage that does not require licensing, however. Such *unlicensed spectrum* is generally intended for use by large numbers of users at very low power, with little likelihood of interference. Early use of unlicensed spectrum was limited to situations where it was assumed that there would be few cases of signal overlap, such as garage door openers and baby monitors. Other services could be manually switched across multiple channels if interference was detected, such as Citizens Band (CB) radio. More recent applications have included adaptive systems that can automatically adjust to accommodate other potentially interfering users, such as WiFi (specified by the IEEE in its 802.11 family of standards). Most of these services are intended for use by the general public (as opposed to commercial operations), largely for their convenience rather than critical or public-safety applications. As a result the requirement for licensing is waived to enable easy adoption by large numbers of users. Unlicensed spectrum is generally not used by broadcasters for their on air operations, with the exception of occasional use of WiFi for first-mile backhaul of content from remote sites via the public Internet when licensed services are not available (see Chapter 11).

White Spaces and Wireless Microphones

Within any given city, a number of TV channels are occupied by broadcast stations, while others are unoccupied, to allow adequate channels for neighboring cities to use without interference. Within those unoccupied channels in any given city, the FCC allows *secondary use* operations to take place on a non-interfering basis. There are two services with such secondary use authorization in the TV band—wireless microphones and *TV White Space* devices.

Wireless microphones, used by broadcasters (see Chapter 10) and others (e.g., theaters and other performing arts venues, musical performers, houses of worship), occupy multiple, narrow channels of spectrum within some or all of a city's unused TV channels, in both the VHF and UHF TV bands. Broadcast users of wireless microphones must be licensed by the FCC. (Other wireless mic users may be licensed, but many are not.) If users of wireless microphones travel to different cities—such as broadcasters on a remote—they may need to adjust their frequencies of operation to not interfere with TV stations in that city.

TV White Space (TVWS), also known as "TV Band" (TVB) devices, are a more recent addition to such secondary usage of TV spectrum. These are wireless data communications systems similar to WiFi. (The term "white spaces" is applied because unused TV-band spectrum is used.) In highly populated areas with many large cities in close proximity to one another, little or no spectrum may be available for such purposes, whereas in less densely populated areas considerable spectrum may be available. Because the UHF TV frequencies used for TVWS have good propagation characteristics for this purpose, the term "super WiFi" is sometimes applied—given that regular WiFi uses higher frequencies (in the 2.5 and 5 GHz regions), which are more easily attenuated by distance and obstructions like walls and floors.

Both licensed and unlicensed operations are permitted in the TVWS environment. Licensed operations are intended for fixed services, intended to be offered by a telecommunications or other service provider, along the lines of a commercial WiFi hotspot. Frequencies of operation are specified in the license. In contrast, personal/portable TVWS services are unlicensed, and are intended for use by general consumers, along the lines of home WiFi access. As such, they share secondary use of the TV spectrum with wireless microphones and similar devices described above. To enable interference-free sharing, all unlicensed TVWS operations must consult an online database (from one of several FCC-approved sources) to determine available frequencies in the usage location. The databases will list all broadcast services that must be permanently protected, along with current temporary uses of wireless microphones and other similar devices by broadcasters and others that have priority over TVWS operations. Broadcasters and other users of wireless microphones and the like must keep the TVWS databases updated regarding their occasional use of these frequencies to prevent interference from TVWS devices.

RF Exposure Regulations

The FCC and other government agencies also enforce regulations intended to protect people from harmful exposure to RF energy. Given that U.S. broadcasters operate some of the highest powered transmitters allowed by the FCC, federal guidelines for human exposure to radio frequency energy apply to all broadcast facilities. Rules vary based on the frequency and power of operation, and include rules regarding the physical containment of transmission sites from intrusion, posting of warning signs, and protection of workers.

Non-Technical Matters

Although they are beyond the scope of this book, for your further study it is important to note that broadcasters are subject to numerous other non-technical and non-spectrum-related regulations enforced by the FCC and other government agencies. Chief among the non-technical FCC rules are those regarding Equal Employment Opportunity (EEO), indecency, and station ownership. The latter stipulates how many stations a single entity can be licensed to operate, both within a single broadcast market and nationwide—with substantially different rules for radio and TV stations in force at this writing. While the FCC's technical rules are fairly stable over time, these non-technical rules tend to be more fluid.

Non-technical rules from other government agencies besides the FCC that affect broadcasters include the compulsory license and other regulations regarding broadcasters' unilateral use of content (e.g., royalties levied on radio stations for their broadcast or streaming of music without direct negotiations with the content's rights holders), enforced by the U.S. Copyright Office, and specialized tax-related rules from the IRS.

Finally, it is important to note that some FCC and other agencies' rules differentiate between commercial and non-commercial broadcasters. These include different protections from interference for non-commercial radio stations in the *reserved band* (a portion of the FM band set aside exclusively for non-commercial operations in most U.S. locations—88.1 to 91.9 MHz), and rules constraining the language that can be used for underwriting or sponsorship announcements on non-commercial radio and television stations.

SPECTRUM AUCTIONS

A significant departure from previous practice was made in 1994, when Congress gave the FCC authority to award spectrum based on auctions, as opposed to comparative hearings or lotteries that had previously been the only methods available. Spectrum "harvested" from legacy users (i.e., made available for various reasons, or no longer required by incumbents) could now be put up for bid by new prospective licensees having the intention of monetizing the use of such spectrum in their businesses, and licensed by the FCC to the highest bidder meeting all necessary qualifications. The key motivator of this change was the burgeoning wireless communications industry (both terrestrial- and satellite-based), in which wireless operators had developed profitable businesses based on the use of spectrum for cellular communications networks offered to customers. Proceeds from such auctions flowed to the U.S. Treasury as revenues analogous to mineral or other usage rights granted to commercial operators on federal lands.

A prominent example of such spectrum auctions involving broadcasters was the disposition of spectrum returned to the FCC at the completion of the DTV transition. Following the shutdown of full-power analog TV channels, and the rearrangement (*"repacking"*) of the UHF TV band, 108 MHz of contiguous spectrum (698–806 MHz, generally referred to as "the 700 MHz band"), which formerly contained 18 TV channels (channels 52–69), was repurposed for wireless telephony and other uses, and most of it was auctioned to wireless service providers. This auction, conducted in 2008, raised approximately $20 billion for the U.S. Treasury. The success of such spectrum auctions in the U.S. led numerous other developed countries around the world to subsequently institute similar processes.

Incentive Auctions

At this writing, the FCC is developing plans for a variant of the process known as *Incentive Auctions*, once again as authorized by Congressional statute. In this case, auctions would be arranged for spectrum that is in use by incumbent licensees, and those licensees would be offered the opportunity to voluntarily divest themselves of spectrum in return for a portion of the proceeds from the auction (hence the "incentive" title), while incumbents who wished to retain their spectrum and continue operations could do so. The incentive auction could also offer possibilities of remuneration for incumbents that wished to remain operational but were willing to share spectrum with one or more other incumbents, or to voluntarily move from their current channel to a less desirable spectral location, or perhaps to tolerate higher interference levels.

To conduct an incentive auction the FCC would engage in a process of arbitrage, by which it would first take bids from the incumbents on what they would "ask" for giving up, sharing or moving their spectrum (called a *reverse auction*), and then iteratively offer the collective spectrum potentially acquired in the reverse auction to prospective buyers (a *forward auction*) until mutual pricing agreement could be reached. After satisfying the accepted reverse-auction bids of incumbents participating in the auction, and covering any costs incurred by the process, net proceeds of the auction would flow to the U.S. Treasury.

At this writing, the first application of this concept is planned for the UHF TV spectrum, with the FCC targeting up to 120 MHz for reallocation via incentive auction to wireless broadband operators. Based on plans announced to date, stations will be offered the opportunity to participate in a reverse auction and provide sealed bids on what they would accept either to give up their channel, to share a channel with another broadcaster, or to move their UHF channel to the VHF TV band. Broadband wireless operators will then participate in a forward auction and bid on spectrum that has potentially been made available by stations that engaged in the reverse auction. At the conclusion of the process, it is assumed that any spectrum vacated will be spread throughout the TV bands across various markets, so the FCC will then have to develop a *band plan* that repacks remaining broadcasters in such a way that all vacated spectrum is shifted into one or more contiguous portions of the UHF band. Therefore many stations that are not participating in the incentive auction will still be affected by it, in being required to relocate their channels to different frequencies. Terms of the auction call for those relocated broadcasters to retain their previous coverage as closely as possible, with a certain amount of the auction's proceeds earmarked for reimbursement of the relocation expenses incurred by such stations.

Given that the process is far more complex than the traditional spectrum auctions, it is unclear at this writing when (or even whether) this auction will be conducted, and how much spectrum ultimately will be reallocated by it. Nevertheless, if successful, it is likely to again become a model for other countries to follow.

CHAPTER 21

Conclusion

This tutorial is now complete. To review, you have learned about the major components of radio and television studios and transmitter sites, their place in the end-to-end broadcast chain, and how the equipment and systems work together to produce the broadcast signal. You have also learned about the basic principles and standards on which broadcasting is based, and something of the role of the FCC in regulating the industry in the United States. You have seen that AM, FM, and TV broadcast facilities have many things in common, but that there are some major differences, particularly in the world of digital broadcasting.

Of course, this is a complex subject and there are many details of the theory, equipment, and systems of broadcasting that have not been covered in this book, but what was included should give you a good understanding of the basic engineering of a radio or television broadcast facility, and of the industry in general. Technologies, of course, will continue to evolve, and disruptions of traditional models will continue to occur, providing an ever-widening range of tools and techniques for making and distributing audio and video programming. Moreover, as the ongoing digital revolution presents consumers with their own new tools to create and distribute content, broadcasters will be driven to reach ever further to provide unique content and services that continue to attract consumers to their offerings. Through its scalability, robustness, and quality, broadcast technology will continue to enable the industry to epitomize the processes of media creation and delivery, while providing unrivaled immediacy and localized relevance to consumers everywhere.

One last caveat: Both creators and consumers of audio or video are only interested in the content that is generated and delivered by a medium's technology, and are generally not enamored by the technology itself. While this may seem self-evident, it bears remembering by those whose work is centered on the technology of such creation and delivery processes. Ideally this technology should "just work" and therefore essentially disappear, allowing the content to reign supreme, and be transparently created by artists and delivered to consumers. As a result, media technologists are often unsung heroes, and their efforts go unappreciated—and this is as it should be. Veteran broadcast engineers understand this well, and know that they typically get attention from colleagues or customers only when something <u>doesn't</u> work as it should. Therefore going unnoticed is a good thing, and a goal to be sought by any broadcast technologist.

Finally, having read this book, it is our sincere hope that you will now feel more comfortable discussing broadcast engineering issues, and in particular, that the information provided will enable you to do your work more effectively.

FURTHER INFORMATION

If you want to explore the topic of broadcast engineering further, many sources of additional information are available. For details on the various aspects of radio and television technology, consult the following resources:

www.nab.org
www.nablabs.org
www.focalpress.com
www.atsc.org
www.smpte.org
www.aes.org
www.ibiquity.com
www.fcc.gov
www.ebu.ch
www.itu.int/en/itu-r

Index

Note: Boldface page numbers refers to figures and tables.

1080i digital HD formats 40, 51, 56
1worldspace Satellite Radio 9
24 frames per second (fps) 31, 32, 51, 52, 142, 253
3D sound systems *see* immersive sound systems
3D TV 53–4, 266
3D video 276
30 frames per second (fps) 32, 38, 51, 55, 142, 166, 253
3:2 pulldown 142
601 video 52
720P digital HD formats 40, 51, 56, 244
8-VSB 10
    modulation 82, 240–2

AAC *see* Advanced Audio Coding
AAF *see* Advanced Authoring Format
AC-3 audio 161, 163, 253–4, 256
acoustic treatment 125
acquisition 140–1
ACR *see* Automatic Content Recognition
active video lines 33
ADC *see* analog-to-digital converter
A/52 Digital Audio Compression Standard 235
A/53 Digital Television Standard 235
Advanced Audio Coding (AAC) 105, 257–8

Advanced Authoring Format (AAF) 156
Advanced Television System Committee (ATSC) 10–11, 235–66
    8-VSB modulation 240–2
    AC-3 audio 253–4
    carriers and channels for DTV 239–40
    channel navigation and branding 239–40
    closed captions 260
    compressed bitstream 242
    content protection (conditional access) 265
    DTV data broadcasting 262–5
    encoder 172, 242, 243
    mobile DTV 12–13
    MPEG-2 compression 246–52
    multicasting 258–60
    multiplexing 258
    NTSC *vs.* 242, 243, 253
    PSIP 260–2
    quality and bit rates 257
    services 265–6
    sidebands and bandwidth 240
    video formats 243–4
AES3 46–7
AES/EBU digital audio 46–7
aggregation sites 223

air chain **90,** 92–3
air-conditioning 126
alternate language audio 173
alternate radio delivery systems
    audio-only service via DTV 227
    audio podcasting 224
    converged receivers 224–6
    hybrid radio 226–7
    internet radio streaming 221–4
    mobile radio APPS 224
alternate television delivery systems
    aggregation sites 270–1
    automatic content recognition 273
    connected television 271
    hybrid TV 274
    internet television streaming and
        downloading 267–8
    mobile television APPS 272
    scaling techniques 268–9
    second screen and social TV 272–3
    social TV 273–4
    TV broadcast services *vs.* online
        content 269–70
    video podcasting 272
AM *see* amplitude modulation
amplitude 25, 77
amplitude modulation (AM)
    antenna systems 290–3
    broadcast facilities 315
    daytimers 302
    IBOC 91, 215–16, 218
    IBOC amplification 288
    IBOC exciters 285
    IBOC system 6
    long wave band 5
    medium frequency 5
    propagation 301–3
    radio 283
    radio amplification 287
    radio exciters 284
    radio wavelength 77
    shortwave radio broadcasting 6
    skywave propagation **302**

amplitude modulation (AM)
    transmission 203–4
    carriers and channels for 204–5
    sidebands and bandwidth 205
    subsidiary communications for
        205–6
analog
    vs. digital 2, 91, 129
    VTRs 145–6
analog audio 26
    sine wave **26**
analog audio signal, single digital sample
    of **42**
analog cable 15
analog color television 31–40
    active video and blanking 33–5
    channel bandwidth 40
    decoding at receiver 38
    field rates 38–9
    film projection 32–3
    frames 31–2
    HD analog video 40
    interlacing, 32
    luminance, chrominance, and
        composite video 37–8
    NTSC 31
    PAL and SECAM 39
    progressive scan 33
    scanning 32, 39
    synchronizing pulses 35–6
    video waveform 36
analog devices 108–10
analog radio 5–6
    AM transmission 5, 203–5
    emissions masks 205–6, **206**
    FM transmission 5, 206–7
    stereo coding 207–11, **208**
    subcarriers 211–12
analog television
    NTSC 9–10
    PAL and SECAM 10
analog-to-digital converter (ADC) 42,
    45–6, 57, 91, 189

analog turntable 108
analog video 29
  high definition 40
analog waveform 41
  monitor 135
ancillary information 232
ancillary systems 164–5
  clocks and timers 118
  intercom and talkback 118–19
  on-air lights and cue lights 119
antennas
  AM 291
  AM IBOC 293
  combined channel 295–6
  directional 294
  FM and TV 293–4
  FM IBOC 295
  multiple bay 294–5
AoIP see Audio over IP
apparatus room 128
A/65 Program and System Information
    Protocol 235, 260
archiving 154
ARIB see Association of Radio Industries
    and Businesses
artist experience 219
aspect ratio 10, 32, 50, 56–7,
    244–6
asset management system 106
Association of Radio Industries and
    Businesses (ARIB) 11, 279
ATSC see Advanced Television System
    Committee
ATSC 2.0 266
ATSC Mobile/Handheld (ATSC M/H)
    174–5
audio 26, 315
  control room 128
  distribution 130
  editing 161
  perceptual coding 222
  podcasting 224
  processing 161–3

production, TV 160
  sweetening 161
  television remote production
    184–5
  waveform 26
audio codec 215
audio compression 215
audio data compression 105
audio delay unit 92, 111–12
audio editing 90
Audio Engineering Society 46
audio files 89
audio-follow-video function 169
audio interconnections 115–16
audio level vs. audio loudness 162
audio mixing consoles 91, 94, **94**
  control surfaces 95
  effects and processing
    units 97
  inputs 95
  monitoring 96–7
  operation 97–8
  outputs 95–6
audio monitoring 96–7
Audio over IP (AoIP) 91–2, 118
audio processing 93
  equipment 113–15, **115**
audio quality 257
audio routing switcher 116
audio storage 90–1
audio-video synchronization
    163–4
audio workstation 103–4
Automatic Content Recognition
    (ACR) 273
automation 88–9
  functions 171
  ingest 165
  systems 105–6
  television 170–1
auto-stereoscopic 3D 53
auxiliary mixes 96
AVC/H.264 252

backward compatible 265
band(s)
   frequency modulation radio,
      VHF 77
   Ku 16
   L-band 7
   S-band 9
   super high frequency 78
   ultra high frequency 78
   very high frequency 5
bandwidth 10, 26
   broadcasting 262
   channel 40, 78
   chrominance signals and 37
   data 63
   FM transmission 207
barn doors 127
BAS *see* Broadcast Auxiliary
      Service
baseband 29, 31, 41
   signals 79
Basic Rate Interface (BRI) 191
BD *see* Blu-ray disc
beam splitter 138
Betacam format (1982) 146
BetacamSP format (1986) 146
Betacam SX (1996) 149
big ugly dishes (BUDs) 16
binary
   bits and bytes 62–3
   data bandwidth 63
   decimal number and **62**
   digits 62
   numbers 61–2
binary values 41
binge viewing 270
bit(s) 44, 58, 61–3
   per second (bps) 63, 253
   rate 45, 57, 216, 218, 241, 244,
      246, 257
   starvation 260
   video user 260
bitrate reduction 58

bitstream 45
   ATSC DTV 284
   compressed 240–2, 258
   distribution 176–7
   recording 176
   serial 45
   server 176
   splicer 176
black burst 38
Blu-ray disc (BD) 67
BRI *see* Basic Rate Interface
broadband wireless 189
Broadcast Auxiliary Service
      (BAS) 309
broadcast facilities, TV 315
broadcasting
   analog radio 5–6
   analog television 9–10
   cable television 15
   definition 5
   digital radio 6–8
   digital television 10–12
   facilities, TV 315
   frequency and power allocations
      308–9
   internet radio and TV 19–20
   satellite radio 8–9
   satellite television 16–17
   telco television 18–19
   terrestrial 22–3
Broadcast Television Systems Committee
      (BTSC) system 208
BUDs *see* big ugly dishes
buffering 251
burnt-in timecode 153
byte(s) 45, 61–3

CA *see* Conditional Access
cable equalizer 117, 158
cable networks 23
cable television
   analog 15
   digital 15–16

CALM Act *see* Commercial Audio
    Loudness Mitigation Act
camcorder 182
camera(s)
    cable 136
    control unit 136
    crane 140
    imaging 137–9, **138**
    news gathering for television 182–3
    scanning technique 37
    slow-motion 185–6
    studio 136–7, **136**
    support 139–40
    television 28
    television remote production 184
camera control unit (CCU) 136
CAP *see* Common Alerting Protocol
caption server 173
carrier(s) 79–80
    analog television 229–30
    digital television 239–40
    for IBOC 214
carrier frequency 203, 205
carrier waves 79, 203, 204
cart machine 108–9
cassette recorders 109
CAT-5 cable 116
cathode ray tube (CRT) 34
    displays 134
CCDs *see* charge-coupled devices
CCU *see* camera control unit
CDN *see* Content Delivery Network
CD-ROM 65
CDs *see* compact discs
CEA *see* Consumer Electronics
    Association
center frequency 78
centralcasting 23, 89, 124–5
central processing unit (CPU) 64
central technical area 128
channel
    back 264
    bandwidth 40, 78

major 239
minor 239
radio band 78
return 264
surround sound 47
channel coding 83
character generator 156
charge-coupled devices (CCDs) 35,
    138–9, **139**
China Mobile Multimedia Broadcasting
    (CMMB) 15
chroma 37
chromakeying system 133
chroma subsampling 49
chrominance 58, 230
    bandwidth 37
    subcarrier 37
circuit switched systems 188
clients 65
cliff effect 213, 242
clocks, ancillary systems 118
closed captions / captioning 172–3
    ATSC digital television
        and 260
    content advisory ratings
        and 232–3
    Federal Communications Commission
        and 232
    vertical blanking interval
        and 260
    as video data 260
CMMB *see* China Mobile Multimedia
    Broadcasting
CMOS technology *see* Complementary
    Metal Oxide Semiconductor
    technology
coaxial cables 15, 47, 58, 68, 116, 158
codecs 10, 58, 71, 189
coded orthogonal frequency
    division multiplexing
    (COFDM) 83, 214
COFDM *see* coded orthogonal
    frequency division multiplexing

color
  bars 37
  black 38
  burst 38
  difference signals 37, 48
  hue 27
  primary 28
  saturation 27
  secondary 28
  subcarrier 37, 230–1
  subsampling 48–9
color filters 28
color subsampling 48–50
color temperature 126
combo studio, radio 87
Commercial Audio Loudness
    Mitigation Act (CALM Act) 163
Common Alerting Protocol (CAP) 113
communication systems 128–9
compact discs (CDs) 44, 67
  players, analog and digital
    connections 102
  ripping 89
  variants 102
Complementary Metal Oxide
    Semiconductor (CMOS)
    technology 139
component video 40
  signals 139
composite, video 37–8, 48
compressed bitstream
    240–2, 258
  distribution 176
compressed links 189
compression
  algorithm 247
  artifacts 253, 259, 260
  audio 255–6
  audio and video data 58–60
  bit rates and 253
computer-based remote control 298–9
computer networks 68–72
computers 64–6

applications 65
automation controlled by 89
clients 65
directories 66
disk drives 66
Internet streaming 71
LANs 68
operating system 65
optical disks 67
personal 64–5
processors 65–6
rack-mount units 65
radio program automation 105–6
security 70–1
servers 65
specialized computers 65
storage 66–8
tape 67
WANs 68
WiFi 69
condenser microphone 99–100, **99**
conditional access (CA) 265
connected cars 225
connected TVs *see* smart TVs
Consumer Electronics Association
    (CEA) 278
  CEA-608 standard 233
  CEA-708 standard 260
content advisory 232–3, 261
Content Delivery Network (CDN) 222, 268
contribution links 187
  for radio 190–2
  for television 193–5
contribution links for radio 190–2
  ISDN line 191
  LAN / WAN 192
  remote pickup 191
  telephone line 190–1
  T1 line 192
contribution links for television 193–5
  microwave 193–4
  satellite 194–5
  video-over-IP 1295

control board 87
Copy Control Descriptor 14
CPU *see* central processing unit
cross-color 231
crosspoint 116
crosspoint switchers 188
CRT *see* cathode ray tube
cue audio 181
cue feed 96, 183
cue lights 119
cue monitoring *see* prefade listen (PFL)
    monitoring
cue tones 108–9
cyclorama 126

DA *see* directional antenna; distribution
    amplifier
DAB *see* Digital Audio Broadcasting
dark fiber 198
DASH *see* Dynamic Adaptive Streaming
    over HTTP
data bandwidth 63
data broadcasting
    digital television 262–5
    equipment 174
    receivers 264–5
data carousel 263
datacasting 10, 262, 263
    service 174
data compression 58–60
Data Over Cable Service Interface
    Specification (DOCSIS) 16
data packets 19
data piping 263
data rate 73
data segments 11
data server 174
data services
    types 263–4
    using all-digital IBOC 219
    using IBOC 219
data, storage *see* storage, data
data streaming 263

data transmission and storage, units for
    63, **64**
DAT recorder *see* digital audio tape
    recorder
DAWs *see* digital audio workstations
daytimers, AM 302
DBS *see* direct broadcast satellite
DCC *see* directed channel change
D9 Digital-S (1995) 149
decoder 250, 251
decoding 38, 250–2
demodulation 80, 233
deviation 207
D1 format (1987) 148
D2 format (1989) 148
D3 format (1991) 148
D5 format (1994) 148
D5 HD (1994) 150
D9 HD (2000) 150
dialog normalization 163
diaphragm 98
digital
    analog *vs.* 2, 91, 129
    components and color subsampling
        48–50
    sideband 215, 217
    signal and noise relationship 47, **48**
    signal robustness 47–8
digital audio 41–7
    concepts, basic 42
    editing 104–5
digital audio and video 41–60
    3D TV 53–4
    AES/EBU digital audio
        standard 46–7
    analog/digital conversion 45–6, 57
    aspect ratio 56–7
    bitstream and bit rate 45
    data compression 58–60
    frame rate 55
    interlacing 56
    lines and pixels 54–5
    mobile formats 52–3

digital audio and video (*Continued*)
quantizing 44
resolution 44–5
sampling 42–4
SD and HD digital video 48–58
signal robustness 47–8
SMPTE serial digital interfaces 58
SPDIF 47
video bit rates 57
Digital Audio Broadcasting (DAB) 7–8
digital audio tape (DAT) recorder 107
digital audio workstations (DAWs)
90, 103–4
Digital Betacam (1993) 148
digital cable 15–16
digital cart 106–7
digital cinema 52
digital component video signals 48–50
digital "islands" 91
digital modulation 240–1
systems 82
Digital Multimedia Broadcasting (DMB) 14
digital radio 41
DAB 7–8
DRM 8
IBOC 6–7
ISDB-TSB 8
Digital Radio Mondiale (DRM) 8
digital record 106–7
digital subscriber line (DSL) 18
digital television (DTV) 41
ATSC 10–11
broadcasting 262–5
DTMB 12
DVB-T 11
ISDB-T 11
master control and emission
encoding **175**
mobile *see* mobile digital
television (DTV)
Digital Terrestrial Multimedia
Broadcasting (DTMB) 12, 279
digital timers 118

digital-to-analog (D/A) converters
42, 91, 189
digital versatile disc (DVD) 67
digital video
formats 50–2
HD 48, 51
SD 48, 50, 51
Digital Video Broadcasting (DVB) 278–9
Digital Video Broadcasting Asynchronous
Serial Interface (DVB-ASI) 176
Digital Video Broadcasting-Cable (DVB-C) 16
Digital Video Broadcasting–Handheld
(DVB-H) 13–14
DVB-SH and DVB-NGH 14
Digital Video Broadcasting–Satellite
(DVB-S) 17
Digital Video Broadcasting–Terrestrial
(DVB-T) 11
digital video effects (DVE) 133–4
digital video recorder (DVR) 57
digital VTRs 147–9, **147**
digital waveform monitors 135
digitizing 155–6
dimmer, lighting 127
diplexers 288–90
direct broadcast satellite (DBS) 17
directed channel change (DCC) 262
directional antenna (DA) 291, 294
directional arrays 291–2
directories 66
direct-to-home (DTH) 17
dish network, DVB-S 17
disk drives 66
distortion 26
distribution amplifier (DA) 117
distribution links 187
distribution system 93
DMB *see* Digital Multimedia Broadcasting
DOCSIS *see* Data Over Cable Service
Interface Specification
Dolby
AC-3 47, 254
digital 254

Dolby A system 110
Dolby B system 110
Dolby E system 161
Dolby® noise-reduction technology 110
Doppler effect 79
Doppler shift *see* Doppler effect
downconversion 166
down stream keyer (DSK) 132
DRM *see* Digital Radio Mondiale
DSK *see* down stream keyer
DSL *see* digital subscriber line
DTH *see* direct-to-home
DTMB *see* Digital Terrestrial Multimedia
    Broadcasting
DTV *see* digital television
dual-sided formats 67
dummy load 289–90
DV and MiniDV (1995) 148
DVB *see* Digital Video Broadcasting
DVB-ASI *see* Digital Video Broadcasting
    Asynchronous Serial Interface
DVB-C *see* Digital Video
    Broadcasting-Cable
DVB-H *see* Digital Video
    Broadcasting–Handheld
DVB- Next-Generation Handheld
    (NGH) 14
DVB-S *see* Digital Video
    Broadcasting–Satellite
DVB-Satellite services to Handhelds
    (SH) 14
DVB-T *see* Digital Video
    Broadcasting–Terrestrial
DVCAM (1996) 149
DVCPRO50 (1998) 149, 182
DVCPRO (1995) 149, 182
DVCPRO HD (2000) 150
DVD *see* digital versatile disc
DVE *see* digital video effects
DVR *see* digital video recorder
Dynamic Adaptive Streaming over HTTP
    (DASH) 269
dynamic equalization 110

dynamic range 257
dynamic ribbon 99

EAS *see* Emergency Alert System
EAV *see* end of active video
EBIF *see* Enhanced TV Binary
    Interchange Format
echo 96
edit controller 131, 152
edit decision list (EDL) 152
edit suites 131
EDL *see* edit decision list
EEO *see* Equal Employment
    Opportunity
effective radiated power (ERP) 292
EHT *see* extra high tension
EITs *see* event information tables
electromagnetic waves 75–7
    description 75
    frequency, bands, and
        channels 77–9
    frequency, wavelength, and amplitude
        76–7
    light as 27, 75
    properties 78–9
    types 75–6
electron gun 34
electronic field production (EFP)
    182, 184
electronic news gathering (ENG)
    83, 148, 182–4
    and EFP 184
electronic newsroom 157
electronic program guide
    (EPG) 10, 261
elementary streams 172
embedded
    audio 47, 159
    systems 66
Emergency Alert System (EAS) 93,
    112–13, **112**
    equipment 170
emissions masks 205–6, **206**

encoder(s)
  ATSC 172, 242, 243
  hardware 72
  MPEG 249, 251, 258
  streaming media 71
encoding, efficient 251–2
encryption 12, 15, 71
end of active video (EAV) 58
ENG *see* electronic news gathering
Enhanced TV Binary Interchange Format
  (EBIF) 16
EPG *see* electronic program guide
Equal Employment Opportunity
  (EEO) 311
equalization 95
equalizer 114, 306
ERP *see* effective radiated power
ether 76
Ethernet 68
ETSI *see* European Telecommunications
  Standards Institute
ETTs *see* extended text tables
Eureka 147 7
European Broadcasting Union 46
European Telecommunications Standards
  Institute (ETSI) 278
event information tables (EITs) 261
exciters 241
  AM IBOC 285
  AM radio 284
  ATSC DTV 285–6
  FM IBOC 285
  FM radio exciters 284
  NTSC TV 285
extended hybrid, FM IBOC 218
extended text tables (ETTs) 261
extra high tension (EHT) 234

FAA *see* Federal Aviation Administration
fader 94, 96
FCC *see* Federal Communications
  Commission

FEC *see* forward error correction
Federal Aviation Administration (FAA) 296
Federal Communications Commission
  (FCC) 22, 203, 205, 307
  antenna height and spectrum
    recovery 238
  ATSC digital television and 235
  broadcast auxiliary spectrum 309
  broadcast frequency and power
    allocations 308–9
  closed captioning information
    and 232
  IBOC 216–20
  incentive auctions 313–14
  licensed *vs.* unlicensed
    operations 310
  mask 214, 215, 217
  non-technical matters 311–12
  RF exposure regulations 311
  rules 229, 232, 237, 244, 256, 308–12
  spectrum auctions 312–13
  vertical blanking interval
    lines and 232
  white spaces and wireless microphones
    310–11
feeder *see* transmission line
fiber-optic cables 47, 68, 158, 185
fiber-optic links 84, 198–9
Fiber to the Node (FTTN) 18
Fiber to the Premises (FTTP) 18
field editing 183–4
fields 32
  "non-integer rates" 39
  rate of 38–9
file-based workflows 130, 160
files and folders 66
file server 103
file transfer 130, 165
film in television 140–2
film projection 32–3
film scanners 142
fingerprinting 273

FiOS 18
firewall 70
first-mile connections 188
flash memory 68
flicker 33
FLO TV *see* MediaFLO
flying spot scanner 141
FM *see* frequency modulation
FOBTV *see* Future of Broadcast
    Television
format conversion 165–6
forward error correction (FEC) 83,
    214–15, 241
fractional frame rates 55
frame(s)
    analog color television and 31–2
    for ATSC digital video 55
    bidirectionally predictive coded 248
    digital audio 46
    intracoded 248
    MPEG 248–50
    NTSC 38–9
    predictive coded 248
    rate 50, 55
frame-compatible ways 54
frame-packing 54
frame synchronizer 159
frequency
    bands, and channels 77–9
    cycle per second 25
    hertz 25
    response 258
    units used 76
frequency coordination 194
frequency modulation (FM) 5, **81**
    broadcast facilities 315
    IBOC 217–18
    IBOC amplification 288
    IBOC exciters 285
    IBOC system 6–7
    propagation 304–5
    radio 283

radio amplification 287–8
    radio exciters 284
    very high frequency band 77
    wireless data 225
frequency modulation (FM)
    transmission 206
    carriers and channels for 207
    deviation, sidebands, and
        bandwidth 207
    stereo signal **210, 212**
frequency response 26
frequency-selective fading 305
frequency swing *see* deviation
FTTN *see* Fiber to the Node
FTTP *see* Fiber to the Premises
Future of Broadcast Television
    (FOBTV) 279

gamma correction 37
gels 127
geostationary satellite 9, 196
ghosting 29, 242
    effect 306
global positioning system (GPS) 118
GOP *see* group of pictures
GPS *see* global positioning system
graphics systems 128, 156–7
graphics tablet 157
ground
    conductivity 292
    radials 292–3
groundwave signal 302
group of pictures (GOP) 249–50
groups 22

hard disk 151
hard disk drives 66
hard disk recorders 89, 103
HD *see* high-definition
HDC™ 119
HDCAM (1997) 150
HDCAM SR (2004) 150

HDC compression 215
HD digital VTRs 150–1
HDMI sticks 271
HD Radio 213
HD radio multicasting *see* In-Band
   On-Channel (IBOC), multicasting
HD-SDI *see* high-definition serial digital
   interface
HDV (2004) 151
headphones 100–1, 225
hearing impaired service 255
helical scan 107, 143–4, **144**
helicopter 183, 193
HEO satellites *see* highly elliptical-orbit
   satellites
hertz (Hz) 25, 76
HEVC *see* High efficiency video coding
Hi8 formats (1989) 146
high-definition (HD) 6–7, 10
   analog video 40
   cameras 139
   digital video 48
   format 129
high-definition serial digital interface
   (HD-SDI) 58
high efficiency video coding
   (HEVC) 277
high-frequency band 6
highly elliptical-orbit (HEO)
   satellites 9
horizontal blanking interval 34, 243
horizontal radiation pattern 292
horizontal sync pulse 36
hue 27
human eye characteristics 28
hybrid IBOC 214–15, 217, 219
hybrid radio 226–7
hybrid radio services 121
hybrid TV 274

iBiquity 213, 219, 220
iBiquity Digital Corporation 6

IBOC *see* In-Band On-Channel
IF *see* intermediate frequency
IFB *see* interruptible foldback
I-frame 248–50, **250**
immersive sound systems 27
IMX (2000) 149
In-Band On-Channel (IBOC) 6–7
   multicasting 120
   operations, facilities for 119–20
In-Band On-Channel (IBOC) digital radio
   213–20
   all-digital phase 214
   amplitude modulation IBOC 215–16
   audio compression 215
   carriers and channels for 214
   data broadcasting 219–20
   frequency modulation IBOC
      217–18
   HD radio standardization 219–20
   hybrid phase 213–14
   modulation and forward error
      correction 214–15
   nighttime operations 215–16
   phased introduction 213–14
   quality and bit rates 216
incandescent 127
Incentive Auctions 313–14
inductive output tubes (IOTs) 287
information technology (IT) 61–73
   binary 61–3
   computer networks 68–72
   computers 63–5
   Internet streaming 71–3
   storage 66–8
ingest 89–90, 165
ingest automation 165
Integrated Services Digital
   Broadcasting—One Segment
   (ISDB 1seg) 14
Integrated Services Digital
   Broadcasting–Terrestrial (ISDB-T)
   11, 279

Integrated Services Digital Broadcasting–Terrestrial Sound Broadcasting (ISDB-TSB) 8
Integrated Services Digital Network (ISDN) 191
   line 197
interactive services 16, 264
interactive streaming 223–4
intercarrier sound 234
intercom system 118, 128
interference 204
interframe coding 248
interlacing 32, 56
interleaving 83
intermediate frequency (IF) 233
International Telecommunications Union (ITU) 52
International Telecommunication Union Radiocommunication Sector (ITU-R) 278
Internet 69–70
   advanced formats 52–3
   TV services 177–8
Internet Protocol (IP) 188, 277
   address 70
   interfaces 166
   unicasting 222
Internet protocol television (IPTV) 19
Internet radio and TV 19–20
   "cord cutting" 20
   OTT services 20
   service implications 20–1
Internet radio operations 121
Internet radio streaming
   aggregation sites 223
   interactive streaming 223–4
   streaming media technology 222–3
Internet service provider (ISP) 222
Internet streaming 71
   data rate 73
   signal quality 73

technology 71–3
   video standards 73
interruptible foldback (IFB) 128
intraframe coding 248
intranet 69
ionosphere 301
IOTs see inductive output tubes
IP see Internet Protocol
IP-based links 189–90
IP-based systems 189
IPTV see Internet protocol television
ISDB 1seg see Integrated Services Digital Broadcasting—One Segment
ISDB-T see Integrated Services Digital Broadcasting–Terrestrial
ISDB-TSB see Integrated Services Digital Broadcasting–Terrestrial Sound Broadcasting
ISDN see Integrated Services Digital Network
ISDT-T 11
isocam 140
ISP see Internet service provider
IT see information technology
ITU see International Telecommunications Union
ITU-R see International Telecommunication Union Radiocommunication Sector

judder 31

keyboard, video monitor and mouse (KVM) hardware 104
Ku-band
   satellite television and 16
   super high frequency 78
KVM hardware see keyboard, video monitor and mouse hardware

LANs see local area networks
last-mile connections 188

lavaliere mics 99
L-band 7
LCD *see* liquid crystal display
leased telephone lines 197
lenses 137
letterboxing 56–7
LFE *see* Low frequency effects
licensed 308
   vs. unlicensed operations 310
light and color 27–8
   electromagnetic wave 27
   filters, color 28
   hue 27
   human eye characteristics 28
   luminance level 27
   primary colors 28
   white 28
lighting grid 126
light modulation 84
lights
   cue 119
   on-air 119
   tally 119
line 27
line 21 35, 232–3
linear faders 95
linear pulse code modulation 46
linear timecode *see* longitudinal
   timecode
line-of-sight propagation 304
links 187–99
   architectures 187–9
   compressed 189
   contribution links for radio 190–2
   contribution links for television 193–5
   IP-based 189–90
   network distribution links for radio and
      television 196–7
   studio-transmitter links for radio and
      television 197–9
lip sync errors 163
liquid crystal display (LCD) 35, 134–5

live assist, radio 89
local area networks (LANs) 68
longitudinal timecode (LTC) 153
longitudinal waves 75
long-playing record 108
long term evolution (LTE) 278
long wave band 5
lossless coding 59
lossy coding 59
loudness 114–15
loudness management 161–3
loudness mismatches 162
loudspeakers 26, 100–1, **101**
   and headphones, subwoofer 254
low frequency effects (LFE) 254
LTC *see* longitudinal timecode
LTE *see* Long term evolution
luma 37
luminaires 127
luminance 37, 48, 58
   color subcarrier and 231
   cross-color and 231
   sampling 48–50
luminance level 27

magnetic data tapes 67
magnetic recording 143
magnetic tape storage 66
main program service (MPS) 218
major channel 239, 260–1
master clock system 118
master control
   network 167
   radio 119
   switcher 168–70, **169**
   television 167–70
master control room 125
Material eXchange Format (MXF) 154
matrixing 209
matte systems 133
MBMS *see* Multimedia broadcast/
   multicast service

MediaFLO 13
medium frequency (MF) 5, 77
medium wave band 5
metadata 120, 174
mezzanine-level distribution 177
MF *see* medium frequency
M formats (1983) 146
microphone(s) 26
   condenser 99–100, **99**
   moving coil 98–9
   news gathering for radio 179–81
   patterns 100
microphone boom 160
microwave 193–4
microwave links 198
M-II formats (1985) 146
minor channel 239, 260–1
mix-effects (M/E) units 132
mixing board, audio 94, **94**
mixing, primary colors 28
mix-minus 110, 181
mobile device, advanced formats 52–3
mobile digital television (DTV) 265–6
   ATSC 12–13
   CMMB 15
   DMB 14
   DVB-H 13–14
   ISDB 1seg 14
   MediaFLO 13
Mobile DTV *see* ATSC Mobile/Handheld
mobile production trucks 185
modem 64
modulation 29
   8-VSB 240–2
   amplitude 5–6, 80, **80**
   carriers 79–80
   COFDM 14, 83
   controlling 115
   digital 240–1
   digital modulation systems 82
   frequency 5, 80–1
   light 84

PSK 82
QPSK 13
   quadrature amplitude 81–2
   sidebands 83–4
   subcarriers 79–80
monitor wall 128
monitor, waveform 36
monochrome TV 37
mono (monophonic) sound 26–7
mono sound, radio studio and 91
motion estimation 248
moving coil microphone 98–9, **98**
Moving Pictures Experts
   Group 246
MPEG-2 compression 105, 246–7, 253
MPS *see* main program service
MSOs *see* multiple systems operators
multicasting 10, 120, 258–60
   Internet streaming 72
   operations 172
multichannel systems 122
Multichannel Video Program Delivery
   (MVPD) systems 177
Multichannel Video Programming
   Distributors (MVPDs) 15
multicore cables 116
multihop link 198
Multimedia broadcast/multicast service
   (MBMS) 278
multipath 83, 241, 304–6
multipath interference 304
multiple systems operators (MSOs) 15
multiplexer 172, 242
multiplexing 205, 258
multitrack systems 122
mult/pool feed 180
MVPDs *see* Multichannel Video
   Programming Distributors
MXF *see* Material eXchange Format

National Radio Systems Committee
   (NRSC) 203, 212

National Television System Committee
(NTSC) 9, 31
active video and blanking 33–5
actual frame and field rates 38–9
channel bandwidth 40
color bars waveform diagram
37–8, **38**
decoding at receiver 38
film projection 32–3
frames 31–2
interlacing 32
luminance, chrominance, and
composite video 37–8
PAL and SECAM 39
progressive scan 33
scanning 32, 39
synchronizing pulses 35–6
video waveform 36
National Television System Committee
(NTSC) analog television
229–34
ATSC *vs.* 242, 243, 253
audio signal 231–2
carriers and channels for 229–30
chrominance information 230
closed captioning and content advisory
ratings 232–3
sidebands and bandwidth 230
vertical blanking interval ancillary
information 232
video signal 230–1
Netflix 270
network 65
cable and satellite 23
distribution system 123
master control 167
release center 123
television **124**
network distribution links for radio and
television 196–7
LAN/WAN 196–7
satellite 196
network switch 70

newsroom 157
Next Generation Broadcasting—Wireless
(NGB-W) 279
next-generation broadcast television
systems
ATSC 1.0. 277
ATSC 3.0 usage scenarios 276–7
audience measurement 280
DTMB 279
DVB 278–9
FOBTV 279
ISDB 279
ITU-R 278
LTE 278
OFDM technology 277
NGB-W *see* Next Generation
Broadcasting—Wireless
nighttime operations 215–16
NLE *see* nonlinear editing
noise 26
noise reduction 109–10
"non-integer rates" 39
nonlinear editing (NLE)
104, 155–6
systems 152
non-program-related data 263
non-real-time (NRT)
broadcasting 266
content 175
nonvolatile memory 66
Notice(s) of Proposed Rulemaking 308
NRSC *see* National Radio Systems
Committee
NRT *see* non-real-time
NTSC *see* National Television System
Committee

OFDM *see* orthogonal frequency-division
multiplexing
omnidirectional microphones 179
on-air automation 170–1
on-air lights 119
"O&Os" stations 22

open-reel recorders *see* reel-to-reel recorders
operating system 65
operations, radio studio 88–91
opportunistic data 262–3
optical
    audio output 46
    disc recording 67
    sound tracks 84
optical disk recording 151
orthogonal frequency-division multiplexing (OFDM) 11
    technology 277
oscillator 287
OTT *see* Over-the-Top
outside broadcast (OB) *see* mobile production trucks
overmodulation 113
Over-the-Top (OTT) 271
    service 20

packets 70
    data 19, 242
packet switched service 188
PAD *see* program-associated data
paint systems 157
PAL *see* Phase Alternating Line
pan-and-tilt head 139, 183
pan-pot 95
passive stereoscopic 3D eyewear 53
patch cord 116
patch panel 116
path loss 79
pay-per-view (PPV) 15
PBXs *see* Private Branch eXchanges
PCM *see* pulse code modulation
peak program meter (PPM) 97
pedestal
    camera 126
    robotic 140
perceptual audio coding 222
perceptual coding 58, 72, 105, 256
persistence of vision 28, 31

personal computers 63–5
PFL monitoring *see* prefade listen monitoring
phase 81
Phase Alternating Line (PAL) 10
phase shift keying (PSK) 82
phosphors 34
photomultiplier 141
pickup cartridge 108
picture 48
picture monitor(s) 28–9, 134–5
pixel 48, 52, 253
Plain Old Telephone Service (POTN) 188
plasma screen 35
platters 66
playback devices 106–7
playlist 89, 171
PMCP *see* Programming Metadata Communication Protocol
podcasting 20, 224
point-to-multipoint paths 188
point-to-point paths 188
polar display 135
portable cameras 137, 139–40
Portable People Meter™ (PPM) 93
portable receiver 181
post-house 124
postproduction 123, 161
potentiometer 94
POTN *see* Plain Old Telephone Service
power amplifier 101, 286
    AM IBOC amplification 288
    AM radio amplification 287
    ATSC DTV Amplification 288
    cooling systems 287
    FM IBOC amplification 288
    FM radio amplification 287–8
    tubes and solid-state devices 287
PPM *see* peak program meter; Portable People Meter™
PPV *see* pay-per-view
preamplier 92

P2 recording format 151
predictive coded frame 248
prefade listen (PFL) monitoring 96
preset-take 169
PRI *see* Primary Rate Interface
primary colors 28
  mixing 28
Primary Rate Interface (PRI) 191
Private Branch eXchanges
  (PBXs) 191
production studio, television **131**
production switcher *see* video, switcher
production trucks 182, 185
profanity delay *see* audio delay unit
professional (PRO) channel 231
Program and Service Data (PSD) 7
Program and System Information
  Protocol (PSIP) 242
  ATSC digital television and 260–2
  directed channel change 262
  electronic program guide 261
  Federal Communications Commission
    and 235
  generator 174
  major and minor channels 260–1
  multiplexing and 258
program-associated data (PAD) 106
program automation software 89
program automation system 105
Programming Metadata Communication
  Protocol (PMCP) 174
program-related data 263
program syndicators 123
progressive download 268
progressive scanning 33, 50, 56
protocol 70
PSD *see* Program and Service Data
PSIP *see* Program and System
  Information Protocol
PSK *see* phase shift keying
Public Switched Telephone Network
  (PSTN) 188
pulse code modulation (PCM) 46, 82, 102

QAM *see* quadrature amplitude
  modulation
QCIF *see* quarter common intermediate
  format
QPSK *see* quadrature phase-shift keying
Quad *see* Two-Inch Quadruplex
quadrature amplitude modulation
  (QAM) 12, 13, 81–2, 205–6,
  215, 231
quadrature phase-shift keying
  (QPSK) 13
quantizing 44
quarter common intermediate format
  (QCIF) 15
quarter video graphics array standard
  (QVGA) 13
QVGA *see* quarter video graphics array
  standard

radio
  frequency 239
  internet radio streaming 221–4
  mobile radio APPS 224
  smartphones 225–6
radio broadcasting
  analog 5–6
  digital 6–8
radio data services 120–1
radio data system (RDS) 212, **212,** 283
radioDNS 226
radio frequency (RF)
  hybrid AM-IBOC **216**
  hybrid FM IBOC **217**
radio frequency (RF) waves 5, 75–84
  bands, channels, and
    frequencies 77–9
  electromagnetic waves 75–7
  over wires and cable 79
radio master control 119
radio news gathering 179–81
radio program automation 105–6
radio receiver 210
radio remote broadcasting 181–2

radio remote productions 181–2
radio studios 87–122
  acoustic treatment 87
  air chain **90,** 92–3
  analog devices 108–10
  analog *vs.* digital 91–2
  ancillary systems 118–19
  AoIP 91–2, 118
  audio delay unit 92, 111–12
  audio mixing consoles 91, 92,
    94–8, **94**
  audio processing equipment
    113–15, **115**
  audio storage 90–1
  automation and live assist 88–9
  CD players 102
  combo studio 87
  data services 120–1
  digital record/playback devices
    106–7
  EAS 112–13, **112**
  editing 90
  hard disk recorders and audio
    workstations 103–5
  IBOC operations 119–20
  ingest 89–90
  Internet operations 121
  live and recorded remote events 122
  loudspeakers and headphones
    100–1, **101**
  master control 119
  microphones 98–100
  multitrack and multichannel
    systems 122
  operations 88–91
  radio program automation 105–6
  remote sources 111
  remote voice-tracking 89
  signal distribution 115–17
  stereo/mono 91
  system considerations 91–4
  telephone hybrids 110–11
  types 87–8

radio wave propagation and broadcast
    regulation 301–14
  AM propagation 301–3
  FCC rules 308–12
  FM propagation 304–5
  IBOC considerations 305–6
  spectrum allocation 307
  spectrum auctions 312–14
  TV VHF AND UHF propagation
    306–7
radio waves 76
RAID *see* redundant array of independent
    disks
RAM *see* random access memory
random access memory (RAM) 64
RBDS *see* United States Radio Broadcast
    Data System
RDS *see* radio data system
reasonable and non-discriminatory
    (RAND) 220
receiver, decoding 38
recorders
  news gathering for radio 179–81
  news gathering for television
    182–3
redundancy 247–8
redundant array of independent disks
    (RAID) 66
reel-to-reel recorders 109
refresh rate 31
reject loads 289–90
remote broadcasting
  cameras and recorders 182–3
  news gathering for radio 179–81
  news gathering for television 182–4
  production, television 184–5
remote dial-up 298
remote pickup units (RPUs) 191
remote sources 111, 168
remote trucks *see* mobile production
    trucks
remote voice-tracking 89
repeaters 9, 297

resolution
  computer screen 55
  digital audio 44–5
  picture 54
  video 50, 253
return channel 16
reverberation 87, 97, 125
reverse auction 313
reverse 3 : 2 pulldown 142
RF *see* radio frequency
RGB (red, green, blue) signal(s)
  37, 40
robotic pedestal 140
router(s), computer network 70
routing switcher 116–17, **117**
RPUs *see* remote pickup units

sampling
  digital audio 42–4
  digital video 48–50
  rate 43
SAP *see* second audio program
SARFT *see* State Administration of Radio,
  Film, and Television
satellite 194–5
  geostationary 9
  HEO 9
  link 89
  networks 23
Satellite Digital Audio Radio Services
  (SDARS) 8
satellite radio
  1worldspace 9
  XM and Sirius 8–9
satellite television 16–17
  digital broadcasting 17
  medium- and low-power services
  16–17
saturation 27
SAV *see* start of active video
S-band 9
SBE *see* Society of Broadcast Engineers
SBR *see* spectral band replication

SBTVD *see* Sistema Brasileiro de
  Televisão Digital
SCA *see* Subsidiary Communications
  Authorization
scanning 32, 39
  progressive 33
SD *see* standard definition
SDARS *see* Satellite Digital Audio Radio
  Services
SDV *see* Switched Digital Video
SECAM *see* Sequential Couleur avec
  Mémoire
secondary colors 28
second audio program (SAP) 208, 234
second screen 272
security, computer 70–1
separate audio program 231
Sequential Couleur avec Mémoire
  (SECAM) 10, 39
serial bitstream 45
serial digital interfaces, SMPTE 58
serializing 48
server-based playout system 167, **168**
servers 65, 104
  file 66
set-top box for IPTV 19
shortwave 6
shot-box 107, 137
shutter, rotating 32
sidebands 83–4, 205, 215, 217, 230, 240
signal
  audio 26
  chain 26
  quality 73
  video 29
signal distribution 115–17, 157–9
  compressed/uncompressed 130
signal-to-noise ratio 110, 206
sine wave 25, 26
  analog audio **26**
single digital sample of analog audio
  signal **42**
single-layer formats 67

Sirius Satellite Radio 8–9
Sistema Brasileiro de Televisão Digital
    (SBTVD) 11
skywave 79, 216, 292, 301–2, **302**
slant track recording *see* helical scan
slow-motion cameras 185–6
smartphones, radio 225–6
smart TVs 19
SMPTE *see* Society of Motion Picture and
    Television Engineers
social TV 273–4
Society of Broadcast Engineers (SBE) 194
Society of Motion Picture and Television
    Engineers (SMPTE) 58, 278
  serial digital interfaces 58
  SMPTE 259M 58
  SMPTE 292M 58
  SMPTE 310M 176
  timecode 152–3
soft lights 127
solid-state flash memory recorder
    179, **180**
solid-state memory 151–2
solid-state recorder 107
solid-state storage 67–8
Sony/Philips Digital Interface 47
sound and audio 26
  mono, stereo, and surround
    sound 26–7
  waves, sound 25–6
sound and vision 25–9
  baseband 29
  light and video 27–9
sound cards, audio 103
sound lock 125
sound stages 125
sound waves 25–6
  amplitude 25
  cycles per second 25
  distortion 26
  frequency 25
  frequency range 25
  noise 26

signal chain 26
sine wave 25
source coding 58
spectral band replication (SBR) 258
spectrum 75
  allocation 307
  auctions 312–13
  BAS 309
  flexible use of 276
SPG *see* sync pulse generator
splatter 205
spotlights 127
squeezeback 169
standard definition (SD) 10
  DTV 50
  format 129
standard definition and high definition
    video 48
  ADCs and DACs 57
  aspect ratio 56–7
  bit rates 57
  components and color subsampling,
    digital 48–50
  formats 50–2
  frame rate 55
  lines and pixels 54–5
  SMPTE serial digital interfaces 58
standards conversion 166
start of active video (SAV) 58
State Administration of Radio, Film, and
    Television (SARFT) 15
stations
  "non-commercial" service 22
  radio and TV 22
statistical multiplexing 259–60
Steadicam™ 140
stereo 130
  coding 207–11, **208**
  generator 209, 283
  multiplex coding **208**
  multiplex matrix **209**
  pilot tone 209
  radio studio and 91

stereoscopic 53
stereo signal generation 209–11
stereo (stereophonic) sound 27
STLs *see* studio-transmitter links
storage, data 66–8
   disk drivers 66
   files and folders 66
   optical disks 67
   solid-state 67–8
   tape 66
   units for 63, **64**
streaming media
   encoder 71
   technology 222–3
studio(s)
   cameras 136–7, **136**
   control room 87–8
   radio *see* radio studios
   site 1
   television *see* television studios
studio-transmitter links (STLs) 1, 92,
      172, 187, 298
studio-transmitter links (STLs)
      for radio and television 197–9
   fiber-optic links 198–9
   ISDN line 197
   leased telephone
      lines 197
   microwave links 198
   T1 and similar lines 197–8
subcarrier(s) 79–80, 211–12
   audio 231–2
   color 37, 230–1
Subsidiary Communications
      Authorization (SCA) 211
subwoofer 101, 254
Super High Frequency 78
supplemental screen 272
surround sound 10, 27, 47, 130, 161,
      254, 255
S-VHS version (1987) 146
Switched Digital Video
      (SDV) 16

switchers 132
switches 70
   computer network 70
symbol, digital 82, 241
synchronizing (sync)
      pulses 35–6
sync pulse generator (SPG) 159
syndicated programs 22
system considerations
   radio 91–4
   television 129–30

talkback system 118–19
tally lights 119
tape
   delay 170
   format 143
   hiss noise 109
   storage 66
TBC *see* timebase corrector
TCP *see* transmission
      control protocol
telco television
   digital optical fiber 18
   digital subscriber line 18
   IPTV 19
telecine 141
telematics 225
telephone hybrids 110–11
telephone line 190–1
teleprompters 136
television
   audio for 160–4
   automation 170–1
   film in 140–2
   master control 167–70
   network **124**
   receiver 28, 38, 237
   remote productions 184–6
   services, Internet 177–8
television news
      gathering 182–4
television studios

advanced programming services 174–5
analog VTRs 145–6
ancillary systems 164–5
audio-video synchronization 163–4
cameras 136–40, **136**
character generators 156
characteristics 125–9
computer-based post-production 134
computer graphics 156–7
control rooms 127–8
digital VTRs 147–9
electronic newsroom 157
file-based workflows 160
hard disk 151
HD digital VTRs 150–1
ingest and conversion 165–6
IP-based infrastructure 166–7
light fixtures and fittings 127
lighting 126
nonlinear editing 155–6
optical disk recording 151
picture and waveform monitoring 134–6
post-production edit suites 131–4
signal delivery to MVPD headends 177
signal distribution 157–9
SMPTE timecode 152–3
solid-state memory 151–2
station and network operations 123–5, **124**
system 130–1, **131**
system considerations 129–30
types 125
video editing 152
video recording 142–52
video servers 153–5
video timing 159–60
terminal adapter 191
timebase corrector (TBC) 144–5
timecode
burnt-in 153
longitudinal 153
signal 143
SMPTE 152–3
vertical interval 153
time delay 170
timers, digital 118
timing of video signals 159–60
T1 line 192, 197–8
total internal reflection 199
towers
guyed 296
lights 296
markings 296
self-supporting 296
wind loading and icing 297
TPO *see* transmitter power output
traffic system 171
transcoding 16, 156, 166
transducers 92
translators 9, 297
transmission channel 32
transmission control protocol (TCP) 70
transmission line 289–90
transmitter power output (TPO) 286
transmitter site facilities 281–99
amplitude modulation antenna systems 290–3
diagram, block **282**
exciters 281, 284–6
frequency modulation and television antennas 293–4
incoming feeds 282
power amplifiers 286
processing equipment 283–4
remote control, transmitter 297–9
towers 296–7
translators and repeaters 297
transmission lines and other equipment 289–90
Transmitter-to-Studio Link (TSL) 197
transport stream 172, 242, 258
transverse wave 75

trick modes 155
trilevel sync 40
tripod, camera 139
tru2way 16
TSL *see* Transmitter-to-Studio Link
tuner 80
turntable 108
TV Band (TVB) devices 310
TV VHF and UHF propagation
    antenna height 306–7
    multipath 306
TV White Space (TVWS) 310
tweeter 101
twisted pair cables 47, 115
Two-Inch Quadruplex (1956) 145
Type C format (1976) 146

UGC *see* user-generated content
UHDTV *see* ultra-high definition
    television
UHF *see* ultra high frequency
ultra-high definition television
    (UHDTV) 52–3
ultra high frequency (UHF) 9, 78
    DAB 7
U-Matic format (1970) 145–6
unicast, Internet streaming 72
Uniform Resource Locator (URL) 223
United States Radio Broadcast Data
    System (RBDS) 212
upconversion 52, 166
URL *see* Uniform Resource Locator
user-generated content
    (UGC) 270
U-verse 18

VBI *see* vertical blanking interval
VCT *see* virtual channel table
vectorscope 135–6
vertical blanking interval (VBI) 33, 35,
    35, 173
    ancillary information 232
    closed captioning and 260

data, discarding unneeded 247
vertical interval timecode
    (VITC) 153
vertical radiation pattern 294
very high frequency (VHF) 78
    band 5, 207
    DAB 7
vestigial sideband (VSB) 230, 240
    modulation 82
VHF *see* very high frequency
VHS version (1976) 146
video 9, 28–9, 173, 315
    3D 276
    active lines 33–4
    analog 29
    ATSC formats 243–4
    bitrate reduction 58
    bit rates 244, 246
    carrier 230
    component 40
    composite 37–8
    compression for recording 147
    converters, analog to digital and digital
        to analog 58–60
    editing 152
    high definition analog 40
    noise 26, 29
    NTSC 284
    patch panels 158
    podcasting 272
    quality 253
    recording 142–52
    routing switcher 158
    signal chain 26
    standards 73
    switcher 131–2, **132**
    timing 159–60
    waveform 36
video distribution amplifier (DA) 158
Video8 formats (1983) 146
Video Home System 146
video-on-demand (VOD) 15, 271
video-over-IP 165, 195

video servers
    archiving 154
    file interchange and mxf 154
    recording formats and interfaces 153–4
    slow and fast motion 155
    types 154
videotape recorders (VTRs) 142–4
    slow and fast motion 145
virtual channel table (VCT) 260
virtual private network
    (VPN) 192
visually impaired service 255
VITC *see* vertical interval timecode
VOD *see* video-on-demand
voice-over 255
Voice over IP (VoIP) 16
voice-tracking 89
VoIP *see* Voice over IP
voltage 26
volume unit (VU) meter 97
VPN *see* virtual private network
VSB *see* vestigial sideband
VTRs *see* videotape recorders

walled garden 225
WANs *see* wide area networks
watermarking, audience measurement 93
waveform 26
    analog 41
    monitors 36, 135
    video 36
wavelength 76, 77
wide area networks (WANs) 68, 188
WiFi 69
wipes 132, 169
wireless microphones 160, 310–11
    for newsgathering 180–1
woofer 101

XDCAM recording format 151
XM Satellite Radio 8–9

YouTube 270
YPbPr components 40

Zenith system 209
zoom lenses 137

Printed in the United States
by Baker & Taylor Publisher Services